MENACE TO THE INNOCENT

INSUBSTANTIAL EXPERT EVIDENCE ENDANGERS INNOCENT PEOPLE ACCUSED OF A CRIME

BY HANS SHERRER

THE JUSTICE INSTITUTE

Menace To The Innocent:
Insubstantial Expert Evidence Endangers Innocent People Accused Of A
Crime

Published by:
The Justice Institute
PO Box 66291
Seattle, WA 98166

http://justicedenied.org
info@justicedenied.org

First posted online, December 2004
First print edition, February 2018

Trade Paperback ISBN: 0-9855033-4-3
Trade Paperback EAN-13: 978-0-9855033-4-5
LCCN: 2018902348

First printing

Cover photo: The horse Clever Hans performing math calculations in the
very early 1900s, and answering by tapping with his right forefoot.

Table of Contents

Author's Note

This book originated from research begun in 2003 for a proposed law review article about prosecutors relying on insubstantial expert evidence to obtain convictions. As it expanded in scope it became too long for a law review article.[1] When it was finished in late 2004 it was intended to be publish it as a book. With the press of other commitments it was uploaded to Justice Denied's website as a temporary expediency.

It is irrelevant that "temporary" situation lasted for more than 13 years: The underlying factors that contribute to prosecutors relying on insubstantial expert evidence remain unchanged. There has not only been no reform of the legal system's structure that allows 'fake' prosecution expert evidence to determine the outcome of many criminal cases, but none is on the horizon.[2] What *has* happened during the past dozen or so years is the discovery many more crime labs have falsified evidence that prosecutors relied on to secure the conviction of thousands of innocent people.

This book's relevance in exposing the long-standing reliance of prosecutors on insubstantial expert evidence is why it is being published in hardcopy in its original form, and it incidentally provides a historical record of the legal system's intractableness as an end driven process to obtain convictions, while giving lip service to the means that are used.

Hans Sherrer
President, The Justice Institute
Editor and Publisher, *Justice Denied: The magazine for the wrongly convicted*
February 15, 2018

---

[1] The almost 1,400 endnotes are due to this book's law review origins.

[2] A non-structural change worth noting is that in 2005 the FBI crime lab discontinuted the examination of bullet lead to link a defendant to a crime, and other crime labs followed its lead. However, the FBI only did that after initially rejecting and refusing to implement the conclusion of a 2004 report by the National Research Counsel that tieing a bullet fragment from a crime scene with a bullet linked to a defendant was a flawed and imprecise procedure. That 214 page report is: *Forensic Analysis: Weighing Bullet Lead Evidence* (Wash., DC: The National Academies Press 2004). The study is available at, www.nap.edu/catalog/10924.html.

This text's discussions of bullet lead evidence is on pages 16-18 and 123. It is prophesized on page 18: "The NAS report may have the impact of providing an avenue for hundreds and perhaps thousands of defendants, an unknown number of whom may have been wrongly convicted, to reopen their cases on the basis of new evidence."

Introduction

We live in an age of magic as a way of life. At least that is how a person who lived 200 years ago could be expected to think of the modern world. In actually, we live in an age of science that to the uninitiated certainly can seem magical. Almost every man-made process we have today that wasn't available 200 years ago is the result of applying scientific principles to varying degrees to achieve the end result.

The quest to solve crimes has not been immune to the application of science. However, this book demonstrates it is not unusual for science to be misapplied, disregarded, or relied on in name only to "solve" a crime and close a case by identifying a person as the culprit. The result is a crime solved by the magical masquerading as science. This situation exists because there to no reliable mechanism to ensure the system isn't gamed by the prosecution's reliance on expert "scientific" evidence that in reality is no more reliable than a confession to being a witch by a person who simply wants to stop being dunked into a pond.

There is generally no scrutiny of crimes "solved" through expert evidence because of the resources necessary to do so, and over 95% of convictions in the U.S. are by a guilty plea that precludes any critical examination of the prosecution's supposedly expert evidence. The overwhelming majority of defendants in this country have limited – if non-existent – financial resources, and public defenders who handle the overwhelming majority of criminal cases have limited budgets, and case load pressure to take the path of least resistance and plead out every case possible.

Consequently, the legal system is structured so that the overwhelming majority of convictions that rely on the soggy foundation of suspect expert evidence – which may in fact be no more stable than quicksand – fall through the cracks into the black hole of a case closed by a plea bargain.

There is relatively little will-power by those within the system to correct this state of affairs. The four primary actors in the legal system's operation – judges, prosecutors, police, and defense lawyers – are integral parts of the assembly line that generates the steady flow of convictions the system depends on for its smooth functioning. The increasing reliance on expert evidence to secure convictions assists to grease the wheels of that system.

The depth of that reliance is demonstrated by how those primary actors exhibit a quasi form of Stockholm Syndrome by their psychological alliance with the use of expert evidence that often is insubstantial and undermines the credibility of the system they are a part of.[3] That psychological state can be called "Expert Syndrome." The way experts are viewed and uncritically relied on masks that their contribution to a case is often no more reliable than the incantation of a witch doctor is to cure an illness or end a drought.

The use of particular expert evidence without full recognition that in an individual case, or in general, it may be unreliable, is irresponsible. The fallibility of expert evidence literally mandates its judicious use. Particularly considering that a consequence of it is an innocent person can spend years in prison, and at worst be executed.

The current system governing the analysis, and presentation of expert evidence is insufficient to protect innocent people from the harmful effects of taking at face value evidence the prosecution purports scientifically proves the defendant's guilt.

This book wouldn't have been written if this was an isolated problem. Quite to the contrary. It is systemic. It is pervasive. It involves fingerprint, DNA, drug, toolmark, and many other types of expert evidence. The legal system functions to facilitate the widespread use of insubstantial expert prosecution evidence to convict untold numbers of innocent people.

The last chapters present substantive out-of-the-box solutions to solve this grave situation. Those meaningful solutions recognize that affixing a feel good band-aid to mask the underlying defects will only allow the system to continue unabated. The innocent can only be protected by withdrawl from relying on a legal framework that doesn't and never will work to protect them.

---

[3] Stockholm syndrome is a condition in which captives develop a psychological alliance with their captors. Unlike hostages, judges, prosecutors, police and defense lawyers chose to develop an alliance with expert evidence that assists in the creation of certainty a suspect/defendant is guilty.

Chapter 1

The Innocent Are Endangered By Insubstantial Expert Evidence

Expert evidence is a valuable prosecution technique used to link an accused person to a crime. However there is a dark side to that technique that can't be ignored: The use of expert evidence as a prosecution tool to falsely implicate an innocent person in a crime. The prevalence of that situation is indicated by irregularities involving prosecution experts that are known to have occurred in the U.S. since at least 1925.[1]

Among irregularities by prosecution experts is substandard work, the withholding of relevant evidence from defense attorneys, reliance on insubstantial or compromised test procedures, and exhibiting bias by slanting evidence reports and testimony to favor the prosecution.[2] Those practices affect criminal prosecutions in a way that is known to be prejudicial to innumerable innocent men and women.[3] Yet they are able to occur with impunity because labeling a testing process as scientific typically assures it of a near mystical reverence.[4] The same is true of people designated as scientific experts.[5] Those experts are interpreters of what is considered to be "the unseen, the indecipherable, or the incomprehensible" to lay persons.[6] A humorous exchange in a cartoon version of Robin Hood captures the essence of the reverence accorded a person identified as a scientific expert:

> Mouse: Who are you?
> Fox: I'm Robin Hood.
> Mouse: You don't look like Robin Hood, but if you say you are then it must be true, because Robin Hood wouldn't tell a lie.[7]

The infallibility attributed to purported scientific tests and the testimony of experts makes it particularly easy for judges, jurors, and defense lawyers to be induced to uncritically accept what are represented as the results of laboratory tests performed on, or analysis of, evidence related to a crime.[8] That attitude is described as the "gee whiz factor" by law Professor Richard Underwood, who credits a person's blind acceptance of expert evidence to "their lack of scientific sophistication and innumeracy."[9] Thus a defendant has little chance of acquittal when a judge or jurors are faced with choosing between circumstantial evidence, or even direct testimony of their innocence, and testimony by a prosecution witness reputed to be an expert.[10] Widely viewed television programs such as *CSI: Crime Scene Investigation* and its spin-offs, and in a previous era *Quincy: M.E.*, contribute to the veneration of forensic experts.[11]

Timothy Durham can swear that isn't theory or conjecture: In spite of 11 alibi witnesses confirming he was in *another state* at the time an 11-year-old girl was raped in Oklahoma, a jury convicted him based on the testimony of a lone expert that his DNA matched the attacker, and that microscopic analysis identified his hair was similar to that of the assailant.[12] Sentenced to 3,000 years in prison, Durham was exonerated when DNA retesting of crime scene evidence excluded him.[13] That proved the prosecution's expert had erred, and his alibi witnesses were right.[14]

Although Durham's case is by no means unique, it exemplifies the gullibility of people to anything labeled as scientific. That naïveté isn't lost on police and prosecutors aware that expert testimony invariably cements a conviction, irrespective of any exonerating non-scientific evidence.[15]

However in a somewhat novel way of looking at scientific evidence, an argument can be made that it doesn't actually exist as evidence at all, because it functions as the prop for the testimony about what it means, which serves as the actual evidence associated with the object/item/thing.[16] This is intuitively supported by evidence being designated as scientific precisely because it is not considered understandable by a juror or judge without being interpreted by the testimony of a person designated as an expert.[17] That evidence only has scientific value to the degree ascribed to it by a person considered to be an expert at interpreting its meaning: The evidence has no intrinsic legal value.[18]

The dependence of the value of scientific evidence on interpretation by an expert takes on a special meaning in criminal cases, since forensic crime laboratories operated by or associated with a police agency work hand in glove with the police agency or prosecutors involved in a case investigation.[19] That relationship exacerbates the handicap faced by a defendant who typically doesn't have the money to independently duplicate the tests performed by a crime lab.[20] Consequently, a defendant invariably relies on the very same forensic results used to substantiate the prosecution's case against him or her.[21] If a defendant does seek to independently test the physical evidence, a response by the prosecution can be that it was all consumed by the crime lab tests.[22]

A crime lab's conflict of interest is compounded by employing workers unevenly trained, lacking certification, and whose work doesn't meet professional standards and lacks objectivity.[23] Under such circumstances, a crime laboratory not following scientific methods can be an important cog in the perfect defendant frame-up loop system.[24] A law enforcement officer can collect or plant false and non-criminatory evidence against a defendant, which is submitted to a crime laboratory whose technician, or technicians, aid the prosecution by falsifying the report about the defendant favorable test result, or who doesn't falsify the report, but testifies falsely about the test's exculpatory nature.[25] Former FBI Director Louis Freeh's expressed the longstanding agency attitude that the FBI Crime Lab's function was "getting results."[26] The defendant frame-up loop is completed when the fake scientific evidence and/or testimony is presented in court as legitimate by the prosecutor.[27]

This process has been going on for generations. Lloyd Miller's case in 1956 is a particularly egregious example of the degree to which a prosecutor will go to use the aura surrounding allegedly scientific evidence to prosecute, convict, and send an innocent man to death row.[28] Miller proclaimed his innocence of murdering a young girl in Canton, Illinois. He was convicted and sentenced to death on the basis of the prosecutors claim that his underwear was found a mile from the murder scene, and a chemist's expert testimony that the clothing had the victim's blood on them.[29]

After refusing to make the clothing available for testing by the defense prior to Miller's trial, the prosecution was compelled by a writ of *habeas corpus* after his conviction to produce them to his legal counsel.[30] It was then proven by an independent laboratory that the red substance on the clothing wasn't blood – it was red paint.[31] It was also proven that the clothing didn't belong to Miller.[32] After the truth about the clothing became known, the prosecutor admitted he knew all along it was smeared with red paint.[33] When the U. S. Supreme Court reversed Miller's conviction in 1967 it plainly stated: "The prosecution deliberately misrepresented the truth. More than 30 years ago this Court held that the Fourteenth Amendment cannot tolerate a state criminal conviction obtained by the knowing use of false evidence."[34] The false expert testimony caused Miller to come within 7-1/2 hours of being executed, and he was eventually granted 10 stays of execution before being exonerated and released after 11 years on death row.[35]

Lloyd Miller's case is a dramatic example of a prosecutor's use of allegedly scientific evidence and testimony to send an innocent man to death row.[36] However it is anything but an isolated occurrence.[37] Furthermore, the ongoing practice of using evidence labeled as scientific to support a conviction when there is otherwise little substantive evidence, is expanding with the increasing use of DNA and other forms of evidence presented as scientific.[38] An example of that beat going on unabated is that 47 years after Miller's conviction, Anthony Bragdon was released in March 2003 after 10 years of wrongful imprisonment.[39] His release followed discovery his conviction of assault with intent to rape was based on the perjured testimony of an FBI crime lab technician that carpet fibers found on the victim's clothing were traceable to Bragdon's apartment.[40]

One of the most troubling aspects of the weighty credibility ascribed by jurors to the probative value of evidence due to its purported scientific testing, and the testimony of a prosecution expert, is it effectively usurps their function as the finders of the facts related to a case.[41] As Professor C.A. J. Coady phrased the problem: "[One] can concede the important, even essential, role of the expert witness . . . [and yet worry about] whether the vastly increased role of experts in the law poses a threat to the proper exercise of the court's arbitral role."[42] If a judge or jurors uncritically accept a prosecution expert's opinion about evidence critical to their determination of guilt, then the finding of guilt was effectively made by the prosecutor successful at getting the jury to hear that damning testimony, and not the jury itself.[43]

Chapter 2

Shoddy Work Is The Norm For Crime Labs

Revelations concerning the shoddy work performed for decades by crime labs are only the tip of the iceberg of the problem nationwide.[44] U.C. Berkeley biochemist and environmental physiologist Benjamin W. Grunbaum has noted that "negligence, incompetence and outright bias" pervaded a cross-section of crime lab reports he examined.[45] Michael Kurland authored a book on forensic evidence describing the situation that continues today: "As for training, many employees of this country's forensic laboratories are inadequately, incompletely, or improperly trained."[46]

Professor Grunbaum also deplored the "orientation of the analyst within the criminal-justice system," whose pro-prosecution bias "makes it difficult to maintain scientific objectivity."[47] He reasoned a "substantial" number of wrongful convictions could result from crime lab practices and the bias of their technicians.[48]

The assistance crime labs are providing in the prosecution of innocent people, whether inadvertent or deliberate, is consistent with the results of the only national proficiency examination of forensic laboratories.[49] The FBI crime lab participated in 18 of the 21 tests, which covered a wide range of the types of physical evidence involved in a criminal investigation – from handwriting analysis to blood type matching.[50] The one constant revealed by the tests was the extraordinarily high rate of errors across the full range of forensic science techniques.[51] Examples are that *51%* of crime labs erroneously identified paint samples, *28%* falsely identified firearms, *71%* made mistakes in analyzing blood samples, *68%* incorrectly matched hair samples, and *36%* failed a soil examination.[52] It is unreasonable to think the results of those test conducted from 1974 to 1977 would be appreciably different today.[53] Particularly considering the consistently high error rate of fingerprint proficiency tests conducted from 1983 to the present.[54]

Although voiceprint analysis was not included in the tests, if it had been the test results likely would have been on the low end of the accuracy scale. A National Academy of Sciences committee rejected voiceprints as an invalidated theory without needing to conduct any proficiency tests.[55] Yet in spite of its lack of a scientific foundation, courts have admitted expert

Menace To The Innocent

voiceprint testimony to assist bolstering the prosecution's case.[56]

The prevalence of inaccurate lab results even extends to testing for the presence of one or more drugs in a urine sample.[57] One example of that is laboratories that participated in a blind test for the presence of one of six drugs, were able to correctly identify the drug 46.5% of the time – which is a lower accuracy rate than could be expected if they had skipped the testing process and simply flipped a coin.[58]

It is consequently not surprising that law Professor Randolph Jonakait expressed the opinion of many when he observed that in regards to the competence of forensic laboratories: "[T]he few disclosed error rates . . . are shockingly high."[59]

The failure of crime labs to reliably arrive at the correct result after what judges, jurors, defense lawyers, interested observers, and the general public are led to believe are incontestably accurate scientific tests, is further confirmed by two different series of annual fingerprint tests involving crime labs across the country.

The first series were conducted by the American Society of Crime Laboratory Directors (ASCLD) each year from 1983 to 1991 as part of its laboratory accreditation program.[60] It was a laboratory test by mail, that involved a dozen or more latent prints that were compared with a number of ten-print cards.[61] A consensus of laboratory personnel had to judge "whether each latent print was scorable, and if scorable, whether it matched a fingerprint on one of the ten-print cards, or could be eliminated as matching none of them."[62] The number of labs participating varied each year from a low of 24 in 1983 to a high of 88 in 1991.[63] If the laboratories consensus response involved only two employees (including both examiners and supervisors), the average individual erroneous identification, i.e., false positive rate during those nine years was 16%, with a high of 30% in 1983 and a low of 10% in four different years.[64] Under the same circumstance, the average over-all error rate (false positives, false negatives, non-answers) on all the tests during those nine years was 41%, with a high of 49% in 1983, and a low of 18% in 1988.[65] If more than two people participated in a lab's response, then the individual rate would be correspondingly higher.

The second series of fingerprint tests have been conducted each year from 1995 to the present by Collaborative Testing Services (CTS), as part of the ASCLD's laboratory accreditation program.[66] It is a test by mail similar to the previous ASCLD test, with the exception that all the latent prints are scorable as either matching or not matching a ten-print card – so the testees know a yes/no response is all that is required for each problem.[67] Mirroring the tests from 1983 to 1991, laboratories submit consensus results, although it does differ in that individual examiners can also participate.[68] Test results are available from 1995 to 2001, and the number of tests submitted yearly varied from 120 to 296.[69] Although CTS does not provide a breakdown of how many labs and individuals take the test, considering the number of labs that participated in the ASCLD tests from 1983 to 1991, it can be estimated that the ratio averages at least two individual examiners for every lab.[70] Based on that ratio of submitted tests, the average individual erroneous identification rate during those seven years was 17%, with a low of 11% in 2001, and a high of 39% in 1995 – which is comparable to the error rate of the ASCLD tests from 1983 to 1991.[71]

Furthermore, not only is the average *over-all* error rate of 36% for those seven years somewhat sobering, but so is the error rate for particular years.[72] In 1995 for example, over half – *56%* – of the tests included at least one error in matching the seven latent fingerprints involved,[73] and *four percent* of the examiners failed to correctly identify *any* of the sample prints.[74] The high error rates for that year were not an anomaly: In 1996 *84%* of the tests included at least one error,[75] and the 1997, 1998 and 1999 tests had over-all erroneous results of 39%, 42% and 38% respectively.[76] In 2001, the last year for which results are available. the over-all error rate was 20%.[77]

The proficiency and certification test results related to erroneous identifications are the error most significant in regards to a criminal case: it indicates how often a lab technician's courtroom testimony about a false positive evaluation implicates a possibly innocent person.[78] The previously cited average individual false positive error rate of 16% in the ASCLD tests, and 17% in the CTS tests only varies by 1% over the 18 year period from 1983 to 2001, and it is 11% for 2001, the most recent year that statistics are available.[79]

The ASCLD and CTS proficiency tests for which results are available extend from 1983 to 2001, and the crime labs and technicians involved had every advantage to provide correct answers: There was no-pressure; no time limit; and, even CTS acknowledges the open book proficiency tests "do not represent the performance accuracy of print examiners in the field."[80] That is particularly true on the test given since 1995, because the testees know all the latent prints either do or do not match a ten-print card – there are no "maybes."[81] Thus on 'gimme' problems under perfect conditions the responding crime labs amassed fingerprint examination error rates that in

Menace To The Innocent

some of the years exceeded the 50% that could be expected if their answers had been chosen by flipping a coin.[82] So two questions that beg to be answered are: How much higher the error rate would be for a proficiency test of fingerprint examiners that simulates real-world conditions;[83] and, what is the real-world error rate.[84]

An indication of the skill level for fingerprint examiners, and hence what their working error rate may be, can somewhat be gleaned from the annual certification of latent print examiners that the International Association for Identification (IAI) has conducted since 1993.[85] The test includes sections on practical knowledge, and includes comparing 15 latent prints with a number of ten-print cards.[86] To qualify to take a certification test, an examiner must have at least two years of full-time experience comparing latent prints, as well as classroom training and practical experience in filing and searching for prints.[87] Although the IAI's certification test are limited to full-time examiners, and *do not* include the full-range of tasks typically performed by examiners, the pass rate form 1993 to 2001 averaged around *50%*.[88] Since to pass an examination a minimum of 80% of the test's fingerprint identifications must be correct, it is known that over-all a minimum of 10.5% of all the test identifications are erroneous, and given averages of a 90% passing score and a 70% failing score, the error rate would be 20%.[89]

Significant crime lab technician error rates also extend to the only known proficiency test of crime lab DNA analysts.[90] In the spring of 1989 the FBI administered an *open* proficiency test to all of its DNA technicians.[91] It was reported by FBI Special Agent Greg Parsons that all the technicians – except for possibly one – failed the test.[92] The reason it is unknown if one technician actually passed the test, is an FBI "supervisor threw away the test results because, according to an FBI agent in the lab: "he feared they would be discoverable" by defense lawyers."[93] Alan Robillard, former head of the lab's DNA Unit, acknowledged to investigators for the Office of the Inspector General (OIG): "that he ordered the results of the proficiency test destroyed, claiming the test was flawed."[94] The only remarkable thing about the test is not the poor showing, which is consistent with every other proficiency test of crime lab technicians conducted during the past four decades, but that *at best*, everyone but one person taking it failed.[95]

Given the known inaccuracy of crime lab tests that is corroborated by proficiency tests, it is reasonable to surmise that if jurors decided on the evidentiary value of prosecution expert testimony by choosing between a short and tall straw while blindfolded, they would achieve a level of accuracy in many cases exceeding that of the crime lab's testing of the evidence testified to by the expert witness.

Chapter 3

Roll Call Of Suspect Crime Labs And Expert Prosecution Witnesses

It would be a matter of concern if irregularities in the operation of a crime lab, the production of unreliable test results, and suspect expert testimony by prosecution witnesses was a unique occurrence. However it isn't. The following 19 brief summaries illustrate how endemic those problems are nationwide in state and federal criminal prosecutions: the FBI Crime Laboratory; Fred Zain – West Virginia's Crime Lab and Bexar County, TX; Orange/Osceola County, Florida's M.E. Office; Oklahoma City P.D.'s Joyce Gilchrist; Phoenix, Arizona Crime Lab; Los Angeles Police Crime Lab; Bexar County, Texas Forensic Science Center; Kansas Bureau of Investigation; Fort Worth, Texas Crime Lab; Montana State Police Crime Lab; Washington State Police Crime Lab; Florida Department of Law Enforcement; Ralph Erdman; Louise Robbins; Michael West; Sandra Anderson; Anthony Pellicano; Chicago P.D. Crime Lab's Pamela Fish; and the Houston Police Department Crime Laboratory.

I – The FBI Crime Laboratory

The FBI crime laboratory is the most prestigious in the country, and its services are only available to law enforcement agencies.[96] Working closely with prosecutors, the lab's testing of suspected evidence is not open to observers for a defendant.[97] Consistent with that secrecy, a court order is necessary for prosecution evidence to be made available to a defendant so it can be tested by an independent forensic laboratory or examined by an independent expert.[98] When appearing in court, an FBI lab technician is accorded a deference that gives their testimony an aura of automatic legitimacy.[99]

That aura of legitimacy results in minimal, if any, scrutiny of the test procedures underlying the testimony of an FBI lab technician. A consequence of the free pass given FBI lab personnel is the ease with which test results and testimony can be fudged or fabricated.[100] Former FBI lab technician Thomas Curran exemplifies the problem and how easy it is for a person engaging in it to go undetected, and then escape meaningful punishment when what they did comes to light.[101] It was reported in *Tainting Evidence: Inside the Scandals at the FBI Crime Lab*:

"Curran had issued reports of blood analyses when "no laboratory tests were done"; had relied on presumptive tests to draw up confirmatory results; and had written up inadequate and deceptive lab reports, ignoring or distorting test results. "The real issue is that he chose to ignore the virtue of integrity and to lie when asked if specific tests were conducted," Cochran's report to the then head of the FBI laboratory, Dr. Briggs White stated. It was an early warning of what could happen at the FBI lab. Tom Curran turned out to have lied repeatedly under oath about his credentials, and his reports were persistently deceptive, yet no one – FBI lab management, defense lawyers, judges – had noticed. When they did, there was no prosecution for perjury."[102]

Thomas Curran's disreputableness did not happen in a vacuum: His actions were fostered by the FBI's culture of focusing on the end of "getting results."[103] As the evidence analysis arm of the federal government's primary investigative agency – that means convictions: Which is why insubstantial expert testimony is tolerated, if not encouraged by prosecutors.[104] There was thus no incentive to change the crime lab's mission of "getting results" after Curran's exposure due to the investigation of a conscientious FBI special agent.[105]

Five years after Curran's exposure, the crime lab's continuing practice of favoring the prosecution was brought up in a report by the GAO.[106] A *Los Angeles Times-Washington Post* wire story in 1997 reported: "The question of potential favoritism in the FBI lab toward prosecutors has been an issue for years. The tainting of crime lab test results was raised in a 1980 General Accounting Office report that criticized the bureau for continuing to staff its laboratory with investigative agents. Other federal laboratories had hired scientists and technicians to examine evidence and thus guarantee impartiality. The bureau rejected the recommendation."[107]

The continuing use of flawed handling and analysis procedures at the FBI laboratory can be attributed to incompetence and/or deliberate subterfuge.[108] The 1980 GAO report indicates that at a minimum it is a combination of the two. A reasonable explanation of why the FBI did not adopt the GAO's common sense recommendations, is provided by its crime lab's role as a key player in framing people by generating inaccurate lab results, falsified reports, and providing insubstantial, if not perjured

Menace To The Innocent

testimony in the *quarter-century* since the GAO report.[109] The manipulation of crime lab technicians even extends to institutionalized racism, by their slanting of a test result and testimony involving a black suspect to ensure it supports the person's guilt.[110]

In 1997 the FBI's crime lab's reliance on questionable practices became front page news when they were publicly revealed by Frederic Whitehurst, a former supervisor of the FBI's explosives lab.[111] Whitehurst became a whistleblower because he wanted to depoliticize the FBI's crime lab, improve the poor quality of its technicians, and rectify the lab's evidence contaminative conditions and deficient handling procedures.[112] His criticism included the lab's handling of the 1993 World Trade Center bombing, and he reported the FBI, especially in high-profile cases, slanted its findings in favor of the prosecution.[113] In typical bureaucratic fashion of killing the messenger, the FBI responded to Whitehurst's disclosures by an act of reprisal: They suspended him.[114]

Underlying Whitehurst's suggested reforms is that professional lab technicians who don't have a history of building cases against criminal suspects are not going to have the same interest in coloring test results to favor the government's case – whether it is done subconsciously or with awareness – as are the poorly trained former investigative agents and other marginally skilled people typically employed by the FBI as technicians.[115]

The FBI lab's advocacy for the prosecution's narrative of a crime led to Whitehurst's first act of whistleblowing eight years earlier.[116] In 1989 Steve Psinakis was prosecuted for allegedly being involved in the illegal shipment of explosives to the Philippines.[117] Whitehurst acted on his conscience by providing the defense with information the prosecution had not disclosed.[118] The information revealed the FBI's expert lab witness was offering an investigative opinion that was not scientifically supported, concerning the "explosives-residue evidence" that underpinned the government's case.[119] The information was pivotal in Psinakis' subsequent acquittal of all charges.[120] After investigating the case years later, the Office of the Inspector General (OIG) found the approach of the FBI's lab agent/technician involved "… represents a fundamental misunderstanding of the role of a forensic scientist."[121]

In the wake of Whitehurst's disclosures, the OIG investigated the FBI's crime lab.[122] The OIG's report released in April 1997 generally confirmed Whitehurst's complaints about the problem of flawed test procedures, the alteration of reports, and misleading and overstated testimony slanted toward supporting the prosecution by FBI lab examiners, who in many cases were unqualified.[123] The following are four of the lab personnel irregularities the OIG discovered:

- Richard Hahn posed as an FBI explosives expert for many years, although he has a degree in chemistry from DePaul University. The OIG report criticized Hahn for "giving scientific opinions that were far outside his area of expertise, and which were unsupported by the evidence."[124]

- Robert Heckman, who has a degree in business administration, was criticized for changing reports, and for his pro-prosecution testimony at the July 1997 trial of BATF undercover agent Carol Howe in Oklahoma that was directly contradictory to his testimony in a previous trial.[125]

- Roger Martz, who has a degree in biology, was the former Chief of the Chemistry and Toxicology Units in spite of not having any training in chemistry or toxicology. Due to his lack of qualifications, the OIG's report recommended that he only be allowed to continue in the lab if he is supervised by, and has all his work reviewed by a qualified scientist.[126]

- Terry Rudolph was found to have not only failed to perform a scientific analysis in numerous cases in which he testified, but that he also lost numerous case files.[127]

The OIG's investigation also found that management of evidence under the control of the lab was nothing short of chaotic, with 8.5% of files missing, empty, or incomplete.[128] Among other things, that deficiency made it impossible to check or replicate test results.[129] However, in spite of its damning conclusions the report was a whitewash of sorts: Since the OIG's investigation was limited to only three of the FBI labs 27 units, the 24 uninspected units were not subject to *any* outside review.[130]

A shadow was again cast on the reliability of the FBI crime lab beginning in April 2003, by public reports concerning three separate situations that demonstrate the lab did not fundamentally alter its operating culture after the release of the Inspector General's report in April 1997.[131] Those involved an FBI lab technician's perjurious testimony during a pretrial hearing in Kentucky murder case about bullet evidence;[132] an FBI lab technician's failure to follow DNA testing protocols in a minimum of 103 cases over at least a two year period;[133] and the conclusion of a report by the National Research Council that the FBI's technique of "data chaining" to link a crime related bullet to a suspect is "flawed and imprecise."[134]

The first of these situations involves the national reporting in April 2003, that FBI lab technician Kathleen Lundy testified during a 2002 pretrial hearing in a Kentucky murder case that "a company melted its own bullet lead until 1996," when she knew at the time "the company actually had stopped in 1986."[135] After testifying truthfully during the trial, she notified her superiors of her false testimony.[136] She said in a sworn affidavit to Justice Department investigators: "I cannot explain why I made the original error in my testimony … nor why, knowing that the testimony was false, I failed to correct it at the time."[137] The United States Attorney declined to prosecute Lundy, however she was charged by Kentucky prosecutors with false swearing, and after pleading guilty, on June 17, 2003, she was given a 90-day suspended sentence and fined $250.[138] In January 2003, before the story broke nationally, the trial judge ruled that Lundy's false pretrial testimony did not prejudice the man convicted of the murder.[139]

The second situation concerns the April 2003 disclosure that FBI lab technician Jacqueline Blake failed to follow protocols by comparing DNA evidence with control samples to ensure the reliability of a DNA analysis.[140] Her conduct began in August 1999, and over a three year period involved DNA evidence in a minimum of 103 cases.[141] After FBI superiors became aware of her activities in mid-2002, at least 29 suspect DNA samples tested by Ms. Blake were identified and removed from the FBI's DNA database.[142] Ms. Blake resigned from the FBI, and the Office of the Inspector General expanded the inquiry into her conduct to encompass investigating the FBI's DNA testing practices in general.[143] Frederic Whitehurst, the whistleblower whose disclosures led to the Inspector General's investigation in 1996-97, noted the disclosures about Ms. Blake's conduct emphasized the need for the FBI's lab to "be subject to independent regulation and inspection."[144] On May 18, 2004, Ms. Blake pled guilty in federal court in Washington D.C. to one count of making false statements on official government reports, although she acknowledged in her plea agreement she knew that any one of her more than 100 false DNA certifications from August 1999 to June 2002 could have been used to identify a suspect in a criminal investigation and influence trial testimony.[145]

The third situation made headlines in November 2003, when it was reported the National Academy of Sciences (NAS) had concluded the FBI had used flawed or imprecise techniques since the early 1960s to match a crime scene bullet to a bullet linked to a suspect.[146] Using a method called "data chaining," FBI technicians have testified during the past five decades in hundreds and perhaps thousands of cases, that a bullet found in the possession of a suspect didn't have to match a crime scene bullet if it could be found to match a third bullet that matched both.[147] The FBI's theory relies on the assumption "that bullets from the same batch of lead share a common chemical fingerprint."[148]

An FBI technician testified in an April 2003 case in Alaska why the practice was so important to the bureau: without "chaining, or something like that, nothing would ever match."[149]

The NAS specifically urged the FBI to stop the practice of using data chaining to declare bullets that *don't* physically match are "analytically indistinguishable."[150] The NAS' report, released in February 2004,[151] also recommends the work and proposed testimony of FBI technicians be peer reviewed "to ensure accuracy and precision."[152]

The NAS' investigation was prompted by the research of retired FBI metallurgist William Tobin.[153] He and his colleagues found "that bullets from the same lead source had different chemical makeups and bullets from different lead sources appeared chemically similar."[154] That finding, which is supported by independent research conducted at Iowa State University, undercuts the foundation of the FBI's reliance on "data chaining," and places it in the realm of junk science.[155] Iowa State University researchers reported: "The fact that two bullets have similar chemical composition may not necessarily mean that both have the same origin. ... The leap from a match to equal origin is enormous and not justified given the available information about bullet lead evidence."[156] The NAS report may have the impact of providing an avenue for hundreds and perhaps thousands of defendants, an unknown number of whom may have been wrongly convicted, to reopen their cases on the basis of new evidence.[157]

Also, in March 2003 it was reported that about 3,000 cases had been identified as possibly being affected by the problems with the FBI lab that were identified in the Inspector General's April 1997 report.[158] The magnitude of the lab's problems is indicated by the report only covering three of the lab's 27 sections.[159] A proportionate extrapolation indicates tens of thousands of criminal cases up until the mid 1990s could involve serious evidentiary problems.[160] The revelations in 2003 about the unabating problems at the lab indicates that untold thousands more cases in the intervening years, including the cases involving bullet "data chaining," could have serious, but as yet undetected evidentiary problems.[161]

Menace To The Innocent

II – Fred Zain – West Virginia's Crime Lab and Bexar County, TX

Fred Zain was a forensic legend. Among his talents was eyesight so acute that he he was able to see "flecks of blood" and "blots of semen" invisible to other examiners.[162] He also achieved test results beyond those of other technicians: Such as his knack of finding "genetic markers" missed by colleagues that were sufficient to implicate a defendant.[163] Fred Zain was thus able to provide crucial evidence prosecutors could not rely on getting from any other lab technician.[164]

Although he worked as a chemist at the West Virginia police crime lab from 1979 to 1989, and then as a supervisor/technician for the Bexar County Forensic Science Center (BCFSC) in San Antonio, Texas, Zain's reputation contributed to his demand as an expert prosecution witness for cases around the country.[165] From Virginia to Hawaii, and from Ohio to Texas he answered the late-night prayers of prosecutors seeking testimony linking a defendant to a crime's physical evidence.[166] It is known his testimony was key to convicting hundreds of rape and/or murder defendants in at least ten states spanning the length and breadth of the country.[167]

The secret to Zain's amazing technical proficiency was discovered in the wake of Glen Woodall's exoneration in 1992 – he faked it.[168] At Woodall's 1987 trial his claim of innocence of sexually assaulting, kidnapping and robbing two women in two separate incidents was overwhelmed by Zain's testimony that his hair was found in each woman's car.[169] Convicted and sentenced to two life terms, Woodall was excluded as the attacker by a DNA test of those same hair samples.[170] Granted a re-trial based on the new evidence, Woodall was acquitted and released after five years of wrongful imprisonment.[171] He was subsequently awarded $1 million as compensation for his ordeal.[172]

A judicial investigation of Zain was initiated following the reversal of Woodall's conviction. The investigation was primarily conducted by the ASCLD, and its conclusions read like a compendium of shady crime lab practices. In its opinion related to the special investigation of Zain, the West Virginia Supreme Court adopted the ASCLD's finding that his misconduct encompassed the following 11 areas:[173]

- Overstating the strength of results.
- Overstating the frequency of genetic matches on individual piece of evidence.
- Misreporting the frequency of genetic matches on multiple pieces of evidence.

- Reporting that multiple items had been tested, when only a single item had been tested.
- Reporting inconclusive results as conclusive.
- Repeatedly altering laboratory records.
- Failing to report conflicting results.
- Failing to conduct or to report conducting additional testing to resolve conflicting results.
- Implying a match with a suspect when testing supported only by a match with the victim.
- Reporting scientifically impossible or improbable results.
- Grouping results to create the erroneous impression that genetic markers had been obtained from all samples tested.

At least 134 cases were identified by the West Virginia investigation as possibly affected by Fred Zain's systematically improper conduct.[174]

After Zain's untoward conduct in West Virginia was publicly disclosed, the Bexar County Forensic Science Center began an internal audit of his work.[175] In November 1993 the audit discovered that contrary to implicating Gilbert Alejandro as Zain had testified at his 1990 rape trial, a test of his DNA had excluded him as the woman's attacker.[176] Eleven months later Alejandro was released after four years of wrongful imprisonment.[177] A Dallas forensic technician who reviewed Zain's work for Bexar County, also "found rampant fraud and falsification. In one case, Zain had testified about blood evidence when no blood had even been found; in other cases he reported performing tests his lab was incapable of doing."[178] Furthermore, a November 22, 1993 internal BCFSC memo outlined that a review of 295 cases Zain worked in 1989 and 1990 revealed that his results in 39 cases didn't match his raw data or worksheets, and in another 29 cases there was no corroborating data – so his conclusions were suspect in at least 63 of 295 cases – over 21%.[179]

As a result of Zain's duplicity, since 1992 at least five men have had their rape and/or murder convictions dependent on his testimony reversed.[180]

Indicted in 1994 in West Virginia for perjury and fabrication of evidence, Zain was able to avoid conviction by expiration of the statute of limitations.[181] He died in December 2002 while awaiting retrial on charges related to perjury and fabricated forensic evidence that resulted in a hung jury in 2001.[182] Zain's prosecution, even absent a conviction, was unusual for a crime lab technician.[183] Some knowledgeable observers have questioned whether Zain's fraudulent courtroom conduct across the country

Menace To The Innocent

even constituted a crime.[184]

III – Orange/Osceola County, Florida Medical Examiners Office

It became public knowledge in September 2002 that in the mid-1990s the medical examiner's office for Central Florida's Orange and Osceola counties was run so slipshod that in hundreds of criminal cases evidence was lost or contaminated, and/or evidence lists and logs were missing.[185] At that time it also came to light that Dr. Shashi Gore was not a board certified forensic pathologist when he was hired in 1996 to run the medical examiner's office and perform autopsies.[186] As a firestorm of controversy swirled around irregularities in his autopsy report concerning the death of 10-week-old Alan Yurko Jr. on November 27, 1997, Dr. Gore announced in the fall of 2003 that he intended to retire in June 2004.[187] Dr. Gore's testimony concerning that report was pivotal to the conviction of baby Alan's father, Alan Yurko, for murdering his son by allegedly shaking him to death.[188] Alan unwaveringly protested his innocence, and after his conviction he assembled an impressive body of evidence substantiating his wrongful conviction, that resulted in the reversal of his conviction on August 27, 2004.[189] One aspect of the insubstantial evidence used against Alan Yurko was the autopsy report Dr. Gore testified about at Alan's trial that was not actually of his son, who was Caucasian, but was the report of a black baby labeled as baby Alan, and who also differed from him by being a smaller, younger child with intact organs, and who had medical ailments he didn't have.[190] In baby Alan's case the medical examiner's office went beyond misplacing or mislabeling a file: It mishandled two children – the autopsied baby Doe, and the still missing baby Alan – who at the time was *the* key evidence at the center of a murder investigation, and whose body was a key to proving his father's innocence.[191] After investigating a complaint filed against Dr. Gore for his conduct in the Yurko case, in February 2004 the Florida State Medical Examiners Board barred him from performing any autopsies until his retirement.[192]

Shoddy lab work by the Pasco County, Florida medical examiner's office also led to an innocent David Long being charged with first-degree murder in the 1998 death of his 7-month-old daughter.[193] Four years later the error was discovered and the charges were dropped.[194] However by then Long had already been jailed, lost his job, spent $100,000 in legal fees and filed for bankruptcy.[195] An indication of the degree to which incompetent, if not outright negligent conduct is tolerated by medical examiners, is that Dr. Marie Hansen, the examiner who erroneously determined the cause of death for Long's daughter, was hired by Dr. Gore to work in the Orange/Osceola

Medical Examiner's Office after Pasco County fired her.[196] He justified her hiring by describing her conduct in Long's case: "It was a small mistake she did."[197]

IV – Joyce Gilchrist – Oklahoma City's Police Lab

In 1986 Jeffrey Pierce was convicted of rape and sentenced to 65 years in prison.[198] His prosecution was based on the expert testimony of Joyce Gilchrist – Oklahoma City's police laboratory chemist – who testified that the attacker's hairs found at the crime scene were "microscopically consistent" with Pierce's.[199]

Fifteen years later Pierce's sperm was excluded by a DNA test as being inconsistent with that of the victim's attacker.[200] Pierce's innocence was corroborated by a lab analysis of the original hair samples used by the prosecution to tie him to the crime.[201] Those tests found that contrary to Gilchrist's testimony, his hair didn't match the attacker's hair.[202] Jeffrey Pierce was released after 15 years of wrongful imprisonment on May 7, 2001.[203]

The FBI's finding about Pierce's case were in a report that described Gilchrist as having "misidentified evidence or given improper courtroom testimony in at least five of eight cases the agency reviewed."[204] The FBI also found her laboratory notes "were often incomplete or inadequate to support the conclusions she testified to."[205] Oklahoma Governor Frank Keating's response was to order an investigation into the over 3,000 felony cases Ms. Gilchrist was involved in from 1980 to 1993.[206]

Although certainly newsworthy, the doubts about Gilchrist's competence reported in the wake of Pierce's exoneration were only news to the public, because they had been expressed for many years by lawyers, other lab technicians, professional organizations, and judges.[207]

James Bednar of the Oklahoma Indigent Defense System observed that for years Gilchrist had testified "with a degree of certainty that did not exist."[208] John Wilson, chief forensic scientist at the regional crime laboratory in Kansas City, Missouri, complained in 1987 to the Southwestern Association of Forensic Scientists (SAFS) that during four trials Joyce Gilchrist provided damning testimony against a defendant that was "not justified by the results of examination."[209] He also said she had "in effect, positively identified the defendant based on the slightest bit of circumstantial evidence."[210] SAFS responded to her conduct by merely verbally admonishing her to "distinguish personal opinion from opinions based upon facts derived from scientific evidence."[211]

The Association of Crime Scene Reconstruction was more severe in its response to her conduct: She was expelled as a member for "unethical behavior."[212]

In 1999 U.S. District Court Judge Ralph Thompson overturned Alfred B. Mitchell's convictions of raping and sodomizing a 21-year-old college student that relied on Gilchrist's testimony about hair and fluid evidence, that he characterized as "terribly misleading, if not false."[213] Judge Thompson also noted that neither Ms. Gilchrist nor the prosecutors on whose behalf she was working, informed Mitchell prior to his trial that DNA tests of the assailants sperm excluded him as the source.[214]

Yet, as of the spring of 2004, Ms. Gilchrist had not been criminally charged as a result of manufacturing testimony out of thin air against innumerable criminal defendants – approximately 400 of who are innocent based on the estimated national 14% rate of wrongful convictions.[215] Less than a handful of those innocent men, however, have had their convictions reversed.

V – Phoenix, Arizona Crime Lab

In July 2003 it was reported that in nine cases, including a murder case, crime lab technicians in Phoenix, Arizona exaggerated the likelihood that a test result implicated the defendant.[216] Interestingly, none of the cases involved DNA testing.[217]

VI – Los Angeles Police Crime Lab

The most thorough lesson in the menace a crime lab can pose to an accused person was presented in the months the O.J. Simpson trial was shown live on national television. Yet that lesson seems to have been lost amidst the widespread speculation about his guilt or innocence that ensued in the wake of the jury's verdict. However all that speculation ignores why it only took the jury two hours to acquit him.[218]

The linchpin of the prosecution's case was its alleged mountain of DNA evidence against O.J.[219] The defense countered that claim by systematically demonstrating the probative value of that alleged evidence was illusory.[220] It was shown that procedural errors during the handling of that evidence, from its lack of isolation and collection at the crime scene, to its transportation to the Los Angeles Police Department (LAPD) crime lab, to its storage and handling at the lab, made the possibility of its contamination so likely that it didn't help the prosecution's case, but fatally undermined it.[221] That doesn't even take into consideration the defense's realistic scenario that DNA evidence was planted, and that the bloody glove allegedly found behind O.J.'s house by the LAPD the morning after the murder wouldn't fit his hand in the courtroom.[222]

The defense was able to demonstrate that the prosecution's vaunted mountain of DNA evidence amounted to nothing more than conclusive proof of the adage: "garbage in, garbage out."[223] Although three labs participated in the testing of the DNA evidence (the California Department of Justice lab in Berkeley, Cellmark Diagnostics, and the LAPD crime lab), all of the DNA evidence was collected, handled, unpacked, stored, repacked for shipping, and in some cases tested by the LAPD.[224] The defense was able to reveal the unreliability of that DNA evidence and testimony related to its prosecution favorable probative value, by demonstrating that contamination was possible, if not likely, at virtually every step of the LAPD's handling of the physical evidence, both prior to and after it the was deposited with the crime lab.[225] During closing arguments the cleanliness of the LAPD's forensic evidence vehicle was compared to a cockroach infested restaurant.[226]

After the verdict, a consensus of observers was the jurors relied on the testimony of the defense's witnesses and arguments discounting the value of the DNA evidence,[227] and without that evidence there was little credible support for a guilty verdict. Although deficiencies in procedures related to the handling and processing of evidence by the LAPD crime lab, that are not untypical for other crime labs, were revealed for all the world to see, nothing substantive is known to have changed in the wake of O.J.'s acquittal.[228] That underscores the only reason the spectacle of the O.J. Simpson trial hasn't been repeated since 1995, is a defendant hasn't been prosecuted in Los Angeles who is well-heeled enough to finance an independent forensic examination of the prosecution's alleged DNA evidence.

VII – Bexar County, Texas Forensic Science Center

Bexar County, Texas is not just notable as one of the counties Fred Zain performed work for.[229] A questionable case he wasn't involved in was the 1993 arson-murder conviction of Sonia Cacy for the November 1991 death of her 76-year old uncle.[230] Sentenced to 99 years in prison, her conviction was solely based on the testimony of a lab technician that traces of gasoline were detected on her uncle's clothing.[231] The two of them were the only occupants of the house that burned down.[232] However, several reputable forensic experts that became involved in her case after her conviction, tested the evidence and found no evidence of gasoline residue.[233] Chemist Gerald Hurst is one of those experts, and he attributed Cacy's conviction to "unrebutted junk science" by the Bexar County crime lab.[234]

Those independent evaluations of the evidence supported both her claim of innocence, and the circumstantial evidence that pointed to her elderly uncle as having accidentally started the fire when he suffered a fatal heart attack while smoking.[235] With mounting national attention being focused on the misjustice of her case, Cacy was paroled in November 1998, after serving more than five years in prison.[236]

A Bexar County case of misjustice that *did* involve Fred Zain was Gilbert Alejandro's 1990 conviction of aggravated sexual assault.[237] Zain testified as Bexar County's chief forensic expert that Alejandro's DNA matched a sample left by the perpetrator on the victim's clothing.[238] It was later discovered during a Bexar County Forensic Science Center (BCFSC) internal audit of Zain's work that he lied during the trial when he implicated Alejandro.[239] At a July 26, 1994 evidentiary hearing, two of the trial jurors, including the foreman, testified that they based their guilty verdict on Zain's testimony, and that in its absence the state would not have met its burden of proving Alejandro guilty beyond a reasonable doubt.[240] On September 21, 1994 Alejandro's indictment was dismissed, and he was subsequently awarded $250,000 in a suit against Bexar County for his four years of wrongful imprisonment.[241] It turned out he had told the truth from the time he was first questioned: At the time of the assault he was at home.[242]

VIII – Kansas Bureau of Investigation

A mislabeled blood sample in the Kansas Bureau of Investigation led to a man being erroneously charged as a serial rapist and murderer in 2002.[243]

IX – Fort Worth, Texas Crime Lab

A senior forensic technician was fired and the operation of the Fort Worth, Texas crime lab was suspended after outside experts discovered discrepancies in cases involving DNA evidence.[244] About 100 cases were involved.[245]

X – Montana State Police Crime Lab

Montana State Police Crime Laboratory Director Arnold Melnikoff's testimony was crucial to 18 year-old Jimmy Ray Bromgard's 1987 conviction of raping an eight-year old girl.[246] The girl said she was only 60-65% sure Bromgard was her attacker after seeing him in a lineup, and at his trial when asked if he was her attacker, she said: "I am not too sure."[247] There were no other witnesses.[248] Bromgard protested his innocence, claiming he was home in bed at the time the attack occurred.[249] In convicting Bromgard, the jury relied on Melnikoff's testimony that there was less than a 1 in 10,000 chance the pubic hairs found on the bed sheets were not his.[250] Ray Bromgard's unrepentant attitude was used by the judge as a reason to enhance his sentence to 40 years in prison.[251] However 15 years later DNA tests conducted on the pubic hair samples excluded him as the girl's attacker,[252] and he was released on Oct. 1, 2002 after 15 years of wrongful imprisonment.[253] After proof of Bromgard's innocence came to light, Melnikoff acknowledged he made up the 1 in 10,000 statistic the jury relied on in convicting him.[254] It is also now known that Melnikoff's sole qualification as a hair analysis was an FBI beginners course in 1975 that didn't include a proficiency test.[255] National publicity about the discrepancy between Melnikoff's testimony and the exonerating test results influenced Montana's governor to order a review of other cases he was involved in.[256]

That review was also warranted by Bromgard's status as the second innocent man known to be convicted of rape based on Melnikoff's suspect testimony.[257] In 1983, Melnikoff linked Chester Bauer's hair sample with that of a rapist.[258] In 1997 Bauer was exonerated and released from prison when it was determined by DNA analysis that the hairs were dissimilar[259] – although Melnikoff's negligence for the error was glossed over by attributing it to a mixing of hairs in the Montana crime lab.[260] A third man convicted of rape on the basis of Melnikoff's faulty testimony, Paul Kordonowy, was exonerated in 2003 after 13 years of wrongful imprisonment.[261]

In 1989 Melnikoff left Montana for Washington state, where he was hired to work at that state's crime lab.[262]

XI – Washington State Patrol Crime Lab

Charges in more than 100 cases were dismissed or never filed that involved a forensic technician in the Washington State Patrol's (WSP) crime lab who admitted stealing heroin evidence from the lab for his personnel use.[263] The technician, Michael Hoover, 'doctored the books' to cover up the thefts.[264] He explained the stolen heroin was used as medication to ease his back pain.[265] After agreeing to a plea deal, Hoover was sentenced in November 2001 to 11 months in jail.[266]

The scandal was the second of the new millennium at the WSP crime lab: One of the lab's senior DNA technicians, Dr. John Brown, resigned in September 2000 while an investigation of his possible mishandling of evidence was being conducted.[267]

In November 2002, the crime lab's operation again came under scrutiny when it was disclosed that lab employee Arnold Melnikoff was under investigation in Montana for conducting tests he had failed to qualify to conduct in Washington.[268] After placing Melnikoff on paid administrative leave in December 2002, the WSP began an internal audit of Melnikoff's work for the lab.[269] The audit report was issued in April 2003, however its contents were not publicly disclosed for almost a year, when the *Seattle Post-Intelligencer* obtained a copy and published a front page expose – *Shadow of Doubt* – on March 13, 2004.[270] The "audit report described Melnikoff's drug-analysis work as "sloppy" and "built around speed and shortcuts."[271] The report's most significant finding was serious questions were raised in 30 of the 100 drug cases he handled between 1999 and 2002, and that the conviction of 22 defendants in 17 cases had been dependent on Melnikoff's suspect handling of evidence, testing procedures, test evaluations, and/or testimony.[272] Evidence in 14 of those cases was recommended for retesting, because Melnikoff's "data was "insufficient" to identify substances."[273] However, the evidence in ten of those cases had been discarded by law enforcement agencies, so retesting was impossible.[274]

As of September 2004, lawyers for the 22 defendants, none of whom had been contacted by prosecutors of the potentially exonerating information in the audit report, were considering various options.[275] The possible responses included moving to vacate the tainted convictions.[276] Melnikoff's pattern of suspect conduct was so pervasive, that Seattle University Law Professor John Strait observed his 30% error rate in tests of physical evidence "wouldn't pass a first-year college chemistry class."[277] Although that error rate may seem somewhat extraordinary, it is consistent with, and in some cases lower than, the error rate in proficiency tests of crime lab personnel nationwide over the past four decades.[278]

Arnold Melnikoff's deficient standard of performance with the WSP crime lab mirrored his conduct as Director of the Montana State Police Crime Laboratory (MSPCL): During his tenure his "powerful, though erroneous testimony swayed" jurors to convict innocent men in at least three rape cases.[279] It was in fact, Melnikoff's insubstantial testimony contributing to the wrongful conviction of Paul Kordonowy in 1990, shortly after he was hired by the WSP, that the agency used to justify his firing on March 23, 2004: "This misconduct now precludes Melnikoff from continued employment with the WSP as a Forensic Scientist 3, as his misconduct as an expert witness was incompetent and was not only inaccurate then and now but was and is contrary to generally accepted scientific principles then and now."[280] However, by firing Melnikoff for his unacceptable conduct related to his work at the MSPCL, the WSP noticeably deflected attention away from his suspect conduct for that agency's crime lab meticulously outlined in the April 2003 audit report.[281]

XII – Florida Department of Law Enforcement

A Florida Department of Law Enforcement technician in Orlando acknowledged falsifying test results intended to measure his proficiency, and the crime labs capability to accurately examine DNA evidence.[282] The department claimed after an *internal* review that the technician's action didn't compromise a test result in any criminal case.[283] However no checks were put in place to prevent a reoccurence of the test falsifications or provide for independent outside verification.[284]

XIII – Ralph Erdmann, M.E. in 42 Texas counties

Working as a contract medical examiner in 42 west Texas counties, Dr. Ralph Erdmann is known to have generated fake autopsy reports of more than 100 bodies he didn't examine.[285] He also falsified toxicology and blood reports in dozens of cases.[286] In many of the cases in which he performed an autopsy, he committed serious errors, including *losing* a man's head.[287] In one case a defendant claimed he didn't touch an 80 year old woman during a robbery of her home, and that she must have died from a heart attack *after* he left.[288] To provide support for the prosecution's otherwise unsupported evidentiary theory that the accused man strangled her, Erdmann testified that she couldn't have died from a heart attack because she had the arteries of a 30-year old woman.[289] To support his contention, Erdmann prepared slides that he testified in court were from the body of the elderly woman.[290] However, he had actually prepared the slides from the arteries of a deceased 30-year old woman.[291] Dr. Erdmann pled no contest in 1992 to charges related to falsifying six autopsies in Dickens, Hockley and Lubbock counties, and he was sentenced to ten years probation.[292]

XIV – Louise Robbins – Forensic Anthropologist

As a footprint examiner extraordinaire, Anthropology Professor Louise Robbins claimed the skill of being able to match a person's footprint impression made on any surface.[293] One of her more remarkable identifications related to testimony that a 3-1/2 million year old footprint "was made by a prehistoric woman who was five and a half months pregnant."[294] Her testimony in what are known as the "Cinderella cases" was

Menace To The Innocent

based on her claim that she was able to match the insole of a suspect's shoe to a footprint.[295] For more than a decade she was a 'hired gun' for prosecutors seeking to cement a conviction, and she appeared as an expert witness in nearly two dozen criminal cases across the country.[296] Although none of her findings were replicated and she never articulated a foundation for her specialized expertise, prosecutors relied on her testimony to seal the conviction of defendants, including an Ohio man whose conviction was quashed after he had spent six years on death row.[297] In another case, charges of capital murder against Stephen Buckley, whose first trial ended in a hung jury in 1985, were dropped when the sole scientific evidence against him, Robbins' footprint testimony, was thoroughly discredited.[298] John Marshall Law School Professor Melvin Lewis tracks thousands of expert witnesses, and he assessed Robbins' work as "complete hogwash."[299] He further explained: "It barely rises to the dignity of nonsense. It's frightening to me that something like that could go as far as it did. Her so-called evidence was so grotesquely ridiculous, it's necessary to say to yourself, if that can get in, what can't?"[300]

XV – Michael West – Forensic Jack of all Trades

Until exposed as a fraud in 1996, Michael West's testimony as a prosecution expert was relied on by prosecutors in ten states.[301] Over 20 capital cases were included among the more than 60 cases for which he provided prosecution favorable testimony.[302] Although he was a board certified dentist by trade, he fulfilled the fantasies of prosecutors by supporting their theory of an alleged crime in areas outside his expertise: In the words of one observer: "West's proclaimed expertise [was] not limited to bite marks. In fact, he … created a comfy niche, mostly as a prosecution expert, matching not only bit marks with teeth, but also wounds with weapons, shoes with footprints, and fingernails with scratches, even spills with stains."[303] West was particularly well known for using "long-wave ultraviolet light and yellow-lensed goggles to study wound patterns on a body."[304] Although that is a standard technique to enhance a deceased person's skin,[305] what set West apart from other forensic examiners was his claim of being able to see lines and marks on a cadaver's skin that no one else could see.[306] A former deputy chief medical examiner who testified against West's findings described his technique as "closer to voodoo or alchemy than science."[307] He also said: "History is full of people who claimed they could see things, from ghosts to UFOs. But claiming it and proving it are two different things."[308]

XVI – Sandra Anderson – Cadaver Finding Dog Trainer

Sandra Anderson was the celebrated trainer and handler of Eagle, described by a fellow trainer as "the most gifted dog in the whole world."[309] Eagle had an uncanny ability to find human remains undetected by any other source, and his talent was in demand all over the world by investigators eager to solve mysterious disappearances.[310] During the ten years they worked together as a team, Eagle and Anderson traveled extensively from their home in Midland, Michigan to such places as Bosnia and Panama searching for mass graves.[311] They also searched for the remains of people on United Flight 93, that crashed in southwestern Pennsylvania on September 11, 2001.[312]

However in April 2002 she was arrested after being observed removing bones from her boot and planting them during a search in Michigan's Huron National Forest, for a woman missing since 1980.[313] Her arrest led to revelations about a number of suspicious 'finds' by Eagle that made it appear he was not a super snoop, but like a scantily clad woman in a magic show, he functioned as an attention distracting prop that allowed Ms. Anderson to complete her illusion of magically finding previously undiscovered evidence that she had actually planted.[314]

Among the suspicious activities that surfaced after her arrest was her work during June 2001, while aiding the Panama Truth Commission to find the remains of victims from 1968 to 1989, of political violence during the reigns of Omar Terrijos and Manuel Noriega.[315] Anthropologist David Martinez claims that when he was working with Anderson he saw a bone fragment fall from one of her socks.[316] The suspicions aroused by Martinez' observation is supported by the determination that at least three sets of bones Eagle found in Panama many miles from each other were from the same body.[317] Martinez was quoted in *The Observer* of London: "I'm thinking maybe this dog is useless and this lady is a liar."[318]

Another suspicious occurrence was that during a search for a missing man in Fulton County, Ohio, Anderson provided police with a neatly severed toe that she said Eagle found in a creek bed.[319] However the 22-year old man's body was found soon after that – with no toes missing.[320] It also came to light after Ms. Anderson's arrest that a blood-stained hacksaw blade Eagle found during a search of Azizul Islam's house for clues related to his missing wife, was actually coated with Anderson's blood.[321] During his October 2000 trial, the jury that convicted him of murdering his wife heard witnesses testify about the blood stained hacksaw blade, and he was

Menace To The Innocent

subsequently sentenced to life without parole.[322]

On August 20, 2003 a ten count federal indictment was issued against Anderson: three counts of obstruction of justice, five counts of falsifying and concealing material facts from federal officers, and two counts of lying to law enforcement officials.[323] The indictment covered two searches in Ohio and five in Michigan.[324] It was reported that as many as 50 criminal cases hinging on evidence provided by Anderson could be reopened.[325]

Sandra Anderson pled guilty on March 10, 2004 to five felonies that involved her planting evidence in at least four cases.[326] The planted evidence included human bones, carpet fibers, a toe, and the bloody saw blade found in Azizul Islam's basement.[327] She also admitted to using her own body fluids to stain coins, a piece of cloth, and the saw blade.[328] On September 28, 2004 she was sentenced to 21 months in prison to be followed by three years of supervised release, and payment of $14,852 in restitution to five law enforcement agencies.[329]

In February 2004, Islam filed a motion for a new trial based on the new evidence that Anderson, an expert prosecution witness at his trial, planted her blood on the hacksaw blade that mention was made of in testimony during his trial.[330] In July 2004 a Wayne County Circuit Court judge ordered a new trial for Islam, and his lawyer told reporters: "Sandra Anderson had been brought in by police departments all around the country to assist them, and as it turns out, she planted it, this hacksaw blade, in [Azizul Islam's] house."[331]

A troubling aspect of the case of Sandra Anderson and her wonder dog Eagle, is that no crime laboratory found anything irregular with the fake bones she found, if they were tested at all, or with her fluids that she planted on objects, including the blood on the saw used to convict Islam, that wasn't his wife's blood, but Anderson's.[332] As for how Anderson signaled to Eagle where the bones were hidden, she may have used nearly imperceptible body movements, which was the method used by Clever Hans, a horse in the early 1900s, to know what answer to provide to complex math problems and other remarkable cognitive feats.[333]

XVII – Anthony Pellicano – Audio Expert Extraordinaire

Anthony Pellicano was known for three decades for two things: As a private eye for Hollywood stars, and as an audio forensic expert – particularly for the prosecution.[334] In dozens of cases, Pellicano testified in support of the prosecution's narrative of the crime, by claiming his enhancement of audio recordings with sophisticated equipment allowed him to understand garbled or faint conversations.[335] Prosecutors typically recruited Pellicano when the FBI or other crime labs found a recording to be unintelligible.[336] Pellicano was also sought after as an expert witness because he was a charismatic man with a persuasive manner who jurors believed.[337]

However his credibility began to unravel when federal prosecutors hired Pellicano to enhance unintelligible recordings of Steve and Marlene Aisenberg, a Tampa, Florida couple suspected of killing their baby.[338] Although the FBI lab was unable to enhance the recordings, Pellicano claimed to hear parts of their conversations, including the alleged comment by Mrs. Aisenberg to her husband: "I hate you. I hate you for what you did to our tiny daughter."[339] He also claimed to have heard her say in the recording: "I tried to save her. She died real bad."[340]

Relying on Pellicano's claims, federal prosecutor's charged the Aisenbergs with making false statements and covering up events related to their daughter's disappearance.[341] When questioned in court about his qualifications as an audio expert, Pellicano "acknowledged that he had no scientific, mathematical or engineering education and did not understand the science underlying his findings."[342] One of the Aisenberg's lawyers observed that Pellicano calling himself an audio expert was as believable as "saying I'm an expert in dentistry because I went out and bought a very fancy drill and I have X-ray equipment."[343]

The prosecution's case collapsed when a federal judge ruled the recordings were unintelligible and therefore testimony about any conversations on them was inadmissible.[344] The judge subsequently awarded $3 million in legal fees to the wrongly accused Aisenbergs.[345]

Anthony Pellicano's career as an expert prosecution audio witness effectively ended on January 23, 2004.[346] That is when he was sentenced to 30 months in federal prison after agreeing to a plea deal stemming from criminal charges filed after federal agents found illegal plastic explosives and two hand grenades in his Los Angeles area office.[347]

XVIII – Pamela Fish – Chicago Police Crime Lab

John Willis was convicted in 1992 of being the man who committed a series of rapes in Chicago beauty shops.[348] Chicago Police crime lab technician Pamela Fish effectively aided the prosecution by testifying Willis could not be excluded as the perpetrator, because tests comparing his DNA to the rapists' were inconclusive.[349] Yet after Willis' conviction, sexual assaults following the distinctive MO of the "beauty shop rapist" resumed.[350] In 1998, six years after Willis' conviction, his lawyer obtained Fish's

Menace To The Innocent

laboratory notes and discovered she had excluded him as the rapist prior to his trial, because he had a different blood type.[351] Based on the new evidence of his innocence, Willis was pardoned by Illinois Governor George Ryan in 1999, and he was released from prison after seven years of wrongful imprisonment.[352]

In October 1986 medical student Lori Roscetti was raped and beaten to death in Chicago.[353] Four youths, Larry Ollins, Calvin Ollins, Marcellius Bradford and Omar Saunders, were charged with the crimes, and their convictions relied heavily on the sperm and hair analysis of the prosecution's expert – Pamela Fish.[354] However her testimony was not consistent between the separate trials of the three defendants who didn't cop to a plea.[355]

Although the Chicago Police Department claimed the crime's evidence had been destroyed, Chicago attorney Kathleen T. Zellner persisted and was able to locate the victim's rape kit, which included the attackers semen on a vaginal swab.[356]

Prosecutors opposed DNA testing to compare the attacker's DNA from the sperm sample compared with that of the four men, based on Fish's "testimony that the men's claims of innocence were baseless."[357] However, after a judge ordered the testing, an independent Canadian forensic lab was selected to analyze the evidence.[358] In May 2001, the lab reported the four men were excluded as the source of the attacker's sperm.[359] The prosecutors then requested that DNA tests be performed on 22 semen stains on Miss Roscetti's clothing, and several other items of evidence, and in September 2001 those test results also excluded the four men.[360] Prosecutors then requested DNA testing of two public hairs and two head hairs believed to be from the rapist/murderer, and in November 2001 those test results also excluded the four men.[361] With the innocence of the four men substantiated by all the physical evidence, their convictions were vacated on December 6, 2001, during a five minute hearing – 15 years after their wrongful convictions.[362] On October 17, 2002, Governor Ryan pardoned the four men, which allowed them to pursue compensation from a state fund.[363]

Subsequent examination of Fish's notes by DNA analyst Dr. Edward Blake revealed that she had known the men were innocent at the time she testified against them, because their blood type didn't match that of Lori Roscetti's killer(s): He was an "O-secretor," while all four of the convicted men were "non-secretors."[364]

In a report on Fish's performance he co-authored with criminologist Alan Keel, Dr. Blake critically assessed Fish's testimony during Larry

Ollins' trial that semen recovered from Roscetti's body could only have been from him: "Fish's representation of her data in this fashion can be viewed only as scientific fraud."[365] The report also concluded in regards to the Roscetti, Willis and other cases: "Ms. Fish misrepresent(ed) the scientific significance of her findings, either directly or by omission." and, "The nature of these errors are such that a reasonable investigator, attorney or fact finder would be misled."[366] Furthermore, the report noted that as a prosecution witness, Fish always "offered the opinion most damaging to the defendant."[367] The report was conducted in conjunction with a lawsuit filed by two other men wrongly convicted on the basis of erroneous testimony by Pamela Fish: Billy Wardell and Donald Reynolds, who were both convicted in 1988 of raping the same woman, sentenced to 55 years in prison, and cleared by DNA tests in 1997 after 9 years of wrongful imprisonment.[368]

Yet in spite of her known track record of repeatedly lying in court under oath, Pamela Fish's career has not suffered: She was promoted to section chief of biochemistry of the Illinois State Police Crime Lab (ISPCL) the same week that John Willis' was released from prison in 1999,[369] and *after* DNA tests exonerated the four men convicted of Lori Roscetti's rape and murder the director of the ISPCL said: "We have no problem with Pam Fish. We have confidence in her work. We've seen nothing to believe otherwise."[370] That stands in stark contrast with Dr. Blake's assessment of Fish's conduct: "The question you have to ask yourself, is whether she was so invested in the prosecution's theory that she bent the results whenever she had the opportunity."[371] Considering the central role Fish's testimony played in the 77 years of wrongful imprisonment inflicted on Billy Wardell, Donald Reynolds, John Willis, Larry Ollins, Omar Saunders, Marcellius Bradford and Calvin Ollins, there is ample support to answer that question in the affirmative.

Attorney Kathleen Zellner's observation helps to explain why – in spite of having no more scientific veracity than the off-the-cuff opinion of a patron in a courthouse coffee shop – Pamela Fish's expert testimony was devastating to those seven men: "Juries are completely trusting of scientific testimony. They think they're hearing impartial science."[372]

XIX – Houston Police Department Crime Lab

In March 2003 Josiah Sutton was released after four years of imprisonment when it was established his conviction was based on a Houston Police Department Crime Lab technician's erroneous testimony that his DNA "definitely" matched that of a car-jacking rapist.[373] That discovery occurred during a wide ranging investigation into the Houston Crime Lab begun in the fall of 2002 by Houston television station KHOU.[374] The station's reports in November 2002 led to the lab's DNA section being shut down in December 2002 pending a full review.[375] Among other things, the station reported the findings of UC Irvine Law Professor William C. Thompson, an expert investigator in finding problems with forensic DNA lab work. [376] After finding serious problems with five of seven cases he reviewed involving DNA testing and testimony, Professor Thompson reported: "It was the worst laboratory work I have seen. The laboratory failed to run essential scientific controls, failed to document their work adequately, and engaged in a variety of practices that create a risk of error."[377] Thompson further found: "[L]aboratory analysts were reporting their results in a misleading manner in written reports and in courtroom testimony. ... They consistently overstated the statistical significance of their findings. ... It appeared the DNA analysts were stretching and distorting their findings to help get a conviction.""[378]

In January 2003 the Texas Department of Public Safety released an investigative report that confirmed Thompson's assessment of the lab's deficiencies.[379] *The Houston Chronicle* picked up on the story and in September 2003 it reported: "None of the analysts who worked in the Houston Police Department's discredited DNA lab were qualified by education and training to do their jobs, based on national standards and a *Houston Chronicle* review of their personnel files."[380] The *Chronicle* also reported in September 2003, that in addition to the Sutton case, significant problems had been found in over 25% of the 49 cases that had been retested since the crime labs deficiencies were unmasked.[381] In November 2003 it was also reported that the lab had lost or destroyed evidence in at least 29 criminal cases, and the Harris County prosecutor announced that the missing evidence would prevent at least 18 defendants from receiving DNA tests.[382]

However the insubstantial work by the lab extended beyond its DNA unit, for example, in two capital murder cases, lab examiner Robert Baldwin did not conclude bullets from the crime scene matched firearms linked to the defendants until after shooting them 25 times and cleaning the barrels.[383] More than 1.300 cases were slated for review due to the disclosures about the crime lab's suspect practices, however the accreditation proposed as a solution for the labs deficiencies would assist more accurate record keeping, but it wouldn't address the lab's proficiency problems.[384]

A Harris County grand jury convened to investigate the crime lab's irregularities expanded its investigation on its own initiative in the Summer of 2003, to include possible wrongdoing by Harris County prosecutors in the generation of erroneous test results, and/or the use of perjurious expert testimony by lab personnel – however as of the fall of 2004 no indictments had been issued.[385]

A new disclosure of crime lab wrongdoing was reported in August 2004: 280 boxes of evidence – that included a fetus, body parts and clothing – involving about 8,000 cases were discovered to have not just been mishandled, but were misfiled by crime lab personnel so their existence was unknown.[386] Following on the heels of uncovered problems with the crime labs DNA, toxicology and ballistics units, the discovery of the misfiled evidence caused the Harris County prosecutor to acknowledge the need for an independent investigation of the labs operation.[387]

Chapter 4

Doctored Tests And Testimony Undermine The Presumption Of Innocence

The notion of a laboratory as a contributor to the search for the truth is undermined in every meaningful way when crime lab personnel conduct a test, write a report, and/or give testimony molded to fit a prosecutor or law enforcement officer's preconceived notions of a person's guilt.[388] That is compounded by the handling of evidence and its exposure to a lab environment that can result in contamination and erroneous results, even when a test is done conscientiously.[389]

What is known about the operating procedures of the FBI and other crime labs would be unacceptable in a high school chemistry class.[390] That has profound implications for society because every person *accused* or merely *suspected* of a crime is legally presumed to be innocent.[391] If a crime laboratory doesn't operate on scientific principles, then its tainted test results undermine the very notion of the presumption of innocence in a way that an innocent person will find hard to refute without the resources for an independent laboratory to conduct scientific tests on the allegedly incriminating evidence.[392]

At a minimum, the presumption of innocence is undermined whenever the law enforcement officials involved in a case don't think they have enough evidence to convict a suspect without a crime lab's assistance in framing the person by fudging or fabricating test results, and then tailoring a technician's courtroom testimony to fit the faked evidence.[393] That attitude reflects the perspective of crime lab personnel that they must provide services that satisfy their prosecution customers.[394] This is just as true when the testimony of a free lance expert, such as Sandra Anderson,[395] Anthony Pellicano,[396] and Michael West,[397] is relied on to support the prosecution's narrative of a case.

The prosecution's overt or unspoken solicitation of inaccurate testing of physical evidence and/or "biased, conflicted, or dishonest" testimony by a crime lab technician or a retained expert flies directly in the face of the theory of American law that a person is considered legally innocent until proven guilty beyond a reasonable doubt.[398] As Justice Brennen clearly stated in the Supreme Court's 1970 case of *In re Winship*:

"The reasonable-doubt standard plays a vital role in the American scheme of criminal procedure. It is a prime instrument for reducing the risk of convictions resting on factual error. The standard provides concrete substance for the presumption of innocence — that bedrock "axiomatic and elementary" principle whose "enforcement lies at the foundation of the administration of our criminal law."[399]

However the presumption of innocence is nothing more than a deceptive mirage when a factual error created by a crime lab, police agency, or outside expert witness is used by prosecutors to frame-up a defendant.[400] The frame-up relying on the artificially created *factual error* is complete when the wrongly convicted person's conviction is affirmed on appeal.[401]

Symbolized by the participation of the FBI and state crime labs, law enforcement's systematic use of false physical evidence to convict people is so sinister that it would make an entertaining B-grade horror movie. It is only too real, however, to every innocent person convicted by the use of such evidence. Compounding that horror are the countless cases where test results or physical evidence exonerating a suspect were destroyed so they couldn't be used to prove the person's innocence.[402] So it is important to recognize "the phenomenon of lost and untested physical evidence" can have as much of a negative impact on an innocent defendant as false testimony by a crime lab technician.[403] Furthermore, it isn't uncommon for evidence that can potentially exonerate a defendant to either not be tested, or if it is, for a favorable test result to be concealed.[404]

Chapter 5

Destruction of Potentially Exonerating Evidence OK With The Supreme Court

The U.S. Supreme Court's 1988 decision in *Arizona v. Youngblood* opened the floodgates to the covert destruction or misplacing of potentially exonerating evidence by police, prosecutors, and crime lab personnel.[405] In *Youngblood* the Court ruled the destruction of potentially exonerating physical evidence by police agencies was not a violation of a person's due process rights if it wasn't done in "bad faith."[406] Larry Youngblood insisted he was innocent of kidnapping and molesting a young boy, and that the destroyed physical evidence could prove it.[407] Yet the Supreme Court reversed an Arizona Court of Appeals ruling that the police's action deprived him of due process and a fair trial.[408] In his dissent, Justice Harry Blackmun wrote:

> The Constitution requires that criminal defendants be provided with a fair trial, not merely a "good faith" try at a fair trial. Respondent here, by what may have been nothing more than police ineptitude, was denied the opportunity to present a full defense. That ineptitude, however, deprived respondent of his guaranteed right to due process of law.
>
> …
>
> It still remains "a fundamental value determination of our society that it is far worse to convict an innocent man than to let a guilty man go free." (citation omitted) The evidence in this case was far from conclusive, and the possibility that the evidence denied to respondent would have exonerated him was not remote. The result is that he was denied a fair trial by the actions of the State, and consequently was denied due process of law.[409]

Twelve years later, in the spring of 2000, a cotton swab was discovered that had been used to collect the attacker's semen from the victimized boy.[410] When the semen was subjected to a DNA test, Youngblood was excluded as being the source.[411] However, being proven innocent was a hollow victory because he had already served his prison sentence.[412] Justice Blackmun's concerns had been proven right and the Supreme Court's majority was clearly mistaken in understating the importance of preserving physical evidence to a person being accorded a fair trial.

Youngblood is still the law of the land in this country even though the rationale underlying it was proven to be legally and scientifically unsound by the subsequent discovery of testable evidence that proved Youngblood's innocence.[413]

It wasn't stated by Justice Blackmun, but since *Youngblood* excuses the outright non-"bad faith" destruction of potentially exonerating evidence with no negative consequences to the prosecution, it was foreseeable that the decision sent the unmistakable signal that every lesser form of mischief related to evidence mishandling – including misplacing and coincidentally "forgetting" it – would become an endemic activity by prosecutors, police, and crime lab technicians.[414]

The degree to which the prosecution's destruction or misplacement of evidence is legally permissible was expanded by the U. S. Supreme Court's unanimous February 2004 decision in *Illinois v. Fisher*.[415] The Court broadened *Youngblood's* reach in *Fisher* by ruling that it includes the destruction of *requested* discovery evidence.[416] The mischief permitted by *Fisher* can be expected to compound the untoward conduct precipitated by *Youngblood*, since the discovery evidence destroyed in *Fisher* was the sole evidence that could have substantiated Gregory Fisher's claim of innocence.[417] That mischief is also compounded by the Court permitting Fisher's conviction to stand even though it was based on a crime lab technician's testimony about what the destroyed evidence was, when there was no verifiable way for the jurors to know if that testimony was truthful.[418] Given what is known about the pro-prosecution bias of crime lab personnel, the judicial green light for police, prosecutors, and crime labs to "non-intentionally" destroy or misplace evidence crucial to a defendant is a bad portent for the innocent: Since prosecution testimony about the probative value of the unavailable evidence cannot be challenged at any point in time by an independent evaluation of that evidence.[419]

Chapter 6

Fingerprint Analysis: Voodoo Palmed Off As Science

The infallibility of fingerprint evidence is akin to a sacred cow in this country.[420] However contrary to that popular belief, the possibility a fingerprint analysis in a given case is erroneous exceeds the probability of a given roll in a crap game.[421] The most serious consequence of that is the evidentiary value of fingerprint testimony in a given case is possibly, if not likely, to be nil.[422]

The dubious value of fingerprint analysis is not a new phenomena, but it is rooted in the very nature of the practice. Insight into the unsubstantiveness of fingerprint analysis comes from understanding that the foundation of fingerprint theory rests on three assumptions:[423]

1) Each person's fingerprints are unique;[424]

2) The uniqueness of each person's fingerprints can infallible be measured;[425]

3) There are people skilled enough to be able to accurately analyze a fingerprint sample with a control sample, and identify that they originated from the same person.[426]

Those three assumptions also underlie the admissibility of fingerprint testimony, because if any one of them isn't true, then not only could it possibly be excludable as non-scientific,[427] but also on the grounds that its prejudicial effect would not be outweighed by its questionable, if not nil probative value.[428]

Fingerprints are unique?

The *first assumption* – that fingerprints are unique – relies on the almost too strange not to be true fact that since 1911, courts in the United States have accepted that assertion *on blind faith* as sufficient to send people to the gallows.[429] Yet while lay persons (including judges and jurors) are generally unaware "the underlying scientific basis of fingerprint individuality has not been rigorously studied or tested,"[430] it is known by forensic professionals[431] and law enforcement officials. An example is the U.S. Department of Justice acknowledged in March 2000 fingerprint individuality has not been scientifically established.[432] However irrespective of a lack of testing, the *theoretical* basis of individuality is suspect.[433] It was noted for example, in a book co-edited by Henry C. Lee, considered one of the world's foremost forensic scientists: "From a statistical viewpoint, the *scientific foundation* for fingerprint individuality is incredibly weak."[434] Thus the widespread belief in the uniqueness of fingerprints is not only scientifically unsupported, but it may be scientifically insupportable.[435]

Fingerprint identification is science?

Doubts about fingerprint singularity are compounded by the methods used to identify what are considered to be distinguishing characteristics of a fingerprint, and that rely on the *second assumption* of fingerprint theory that their identification is an exact science.[436] There were differing methods of physically identifying a person during the mid-to-late 19th century.[437] However, there was no scientific basis established for the accuracy of any of them.[438]

Englishman Francis Galton led the way in the consideration of fingerprints as an identification method, by developing "Galton points" as a technique of comparing fingertips ridges between two samples.[439] Galton's techniques, however, did not automatically inspire confidence that they had a sound foundation: It is observed in *Suspect Identities* that the British Home Office rejected the use of fingerprints for identification purposes in 1894, because "there was no reason to resort to an unproven technology like fingerprints."[440] In spite of such doubts about its scientific veracity, the "Galton point" method of fingerprint identification[441] eventually enjoyed widespread adoption beginning in 1897 in India, because it was considered an efficient way to record a physical characteristic that could be classified, cataloged and retrieved with relative ease.[442]

Thus the adoption of fingerprints as an identification method was driven by bureaucrats who embraced it as meeting their work requirements, based on the scientifically unsubstantiated conjecture of its proponents about its reliability.[443] That historical oddity is still relevant: The selection of fingerprint analysis out of convenience continues to be a primary, although typically unstated motivating force behind the defense of its use today on grounds of common practice in this country for almost 100 years.[444] That justification is substituted for the lack of it having been proven to have a scientific basis.[445]

Fingerprint examinations are accurate?

Doubts about the scientific foundation of fingerprint analysis are

Menace To The Innocent

compounded by the dependence of the process on a human interpreter's subjective evaluation, which relies on the *third assumption* of fingerprint theory that examiners have special skills enabling them to pinpoint tell-tale fingerprint identifiers.[446] In 1892, Galton recognized the inherently subjective nature of fingerprint examination when he wrote in *Finger Prints*: "A complex pattern [like fingerprints] is capable of suggesting various readings, as the figuring on a wall-paper may suggest a variety of forms and faces to those who have such fancies."[447] Thus one of the people most responsible for the use of fingerprinting as an identification mechanism acknowledged it was not in the nature of a precise scientific endeavor.[448] It is not without reason that critics of fingerprint analysis have compared it to pseudo-sciences such as palmistry and handwriting analysis.[449]

I – The Black Art of Fingerprint Analysis

In spite of being allowed in courtrooms as presumably scientific, numerous authorities have recognized as Galton did in 1892, that fingerprint analysis is dependent on the subjective opinion of each examiner.[450] It does not, for instance, involve an objective process with duplicatable results consistent with the requirements of the scientific method.[451] It was observed in *Fingerprints: What They Can & Cannot Do* that the identification process "is a subjective determination by the examiner based on the individual's training, experience, and abilities as his skills are applied in the evaluation of any particular print."[452] That is consistent with forensic scientist David Stoney's observation in a legal practice guide that: "the criteria for absolute identification in fingerprint work are subjective and ill-defined. They are the products of probabilistic intuitions widely shared among fingerprint examiners, not of scientific research."[453] Dr. Stoney testified in *U.S. v. Mitchell* regarding fingerprint analysis: "By not scientific, I mean that there is not an objective standard that has been tested; nor is there a subjective process that has been objectively tested. It is the essential feature of a scientific process that there be something to test, that when that something is tested the test is capable of showing it to be false."[454] Dr Stoney's observation that the subjectivism of fingerprint analysis excludes it as a scientific endeavor, is a modern update of concerns about the soundness of its foundation going back to the 19th century.[455]

In the years following 1910, when fingerprint evidence was first used to convict a person in the U.S., its proponents felt compelled to try to distinguish the practice from "palmistry" and other pseudo-sciences.[456] In today's parlance, early fingerprint practitioners had to overcome the perception they were the equivalent of a tarot card reader or a psychic reader similar to television's Miss Cleo.[457] Yet such efforts can never be wholly successful, since as was recognized by two renowned authorities in *Advances in Fingerprint Technology* (2001), fingerprint identification depends on "human intuition."[458] Hunches, guesswork and supposition are essential tools of the trade for a fingerprint examiner.[459]

Thus allusions to fingerprint analysis as being a scientific endeavor serve to give it respectability by masking that it is akin to a 'black art.'[460]

II – The Disparity Between Latent and Controlled Fingerprint Samples

The essence of a fingerprint examination in a criminal case is a suspect's full set of ten fingerprints are compared for similarities with one or more latent fingerprints. The suspect's fingerprints are obtained under quasi-controlled conditions, and the latent print(s) are recovered from a location such as a crime scene.[461] The former invariably have significantly sharper characteristics than the latter.[462] The reason for the clarity difference is a print is a representation of the ridges of skin on a fingertip and what is on those ridges, and the clarity of an impression is dependent on the quality of the surface, and the circumstances under which it is made.[463] In a quasi-controlled setting such as a police station, the entire friction-ridge surface of all ten fingers can be inked and rolled onto a highly receptive surface to make a clearly visible impression.[464] In contrast the quality of a latent print found at a remote location, such as a crime scene, is degraded from being created in an imperfect environment under imperfect conditions:[465] They are subject to such vagaries as smudging; variations in pressure; only a small portion of a print is typically recoverable;[466] the surface they are found on can be uneven, porous and/or soiled with sweat, water, dirt, blood, oil or some other substance; and they can be cross-contaminated with other prints or an obfuscating agent.[467] All of those factors affect the clarity of the latent print, and any one of them can reduce the likelihood it will match a print obtained under controlled conditions – even if the two prints are identical.[468] That is why examiners typically refer to being "comfortable" prints are sufficiently similar to declare a match – not that they are convinced they are exactly the same.[469]

Further contributing to the difference in quality between controlled and latent prints is the fact that the latter are invisible to the naked eye.[470] At their most faint, a latent print is left on a surface by the perspiration exuded by minute sweat pores that adheres to ridges on a person's fingertips.[471] Contaminants on the ridges, along with perspiration and body oils can enhance an impression left on a surface, or they can even be partially transferred to the surface.[472]

Due to their invisibility, a latent print must be treated with "some kind of powder, chemical, or electronic processing, or enhancement," to transform it into being visible.[473] So all latent fingerprints are by definition "filtered" by the process used to make them visible.[474] Furthermore, the filtering process can cause distortion in the latent print it makes visible, because it is not a neutral procedure, and can enhance, obscure, or create features of the invisible (latent) fingerprint.[475] Those factors are compounded by elastic deformity of a print caused by variations in the pressure of a fingertip on a surface.[476] Furthermore, if a surface that is touched has a receptive coating such as dust, undried blood, wet paint, etc., then regardless of what is on the fingers ridges, a deformed impression can be left on that surface.[477]

III – Every Person's Fingerprint Matches On Some Level

The degraded features and small size of a latent print – generally only $1/5^{th}$ the size of a quasi-controlled print[478] – particularly undermines its value as an identifying medium, and greatly increase the probability of a false match with a print sample obtained under quasi-controlled conditions.[479] The reason for this is explained by an important phenomena: "The smaller the sample that is compared to a primary object, the greater the likelihood there is of a correspondence."[480] Consequently, the smaller the area covered by a fingerprint sample, the greater the likelihood a match will be made when it is compared to the complete print samples of other people.[481] It is within the realm of consideration that given a small enough area of comparison, every person on earth with ten identifiable fingerprints could be matched to each other.[482] An article that appeared in the Spring 2003 issue of *Justice Denied* magazine, *A Printer Looks At Fingerprints*, explains this phenomenon in plain terms:

"An illustration of this principle can be made by supposing that some play of Shakespeare be compared against [this article], in its totality. Obviously they will be entirely dissimilar. Now compare some isolated short sentence from this journal to every sentence in Shakespeare: there may or may not be two that are exactly the same, but at least the odds will now be tremendously greater in favor of a correspondence. Finally, if you selected some *pair of words* and then hunted all through the Bard's work for them, odds would approach 85% or higher that you'd find that identical couple. At a single word the odds would rise still farther, and at individual letters of the alphabet the odds would, of course, be precisely 100%."[483]

This mathematically reducible peculiarity is compounded by salvaging small fragments of a latent print as evidence, through the use of computerized enhancement techniques.[484] An implication of this principle is that even if at some point in the future every person's full-set of ten fingerprints was absolutely proven to be unique, it is likely that multiple sections of every person's fingerprints are identical at some scale of comparison, to a section of any number of other people's fingerprints.[485] Consequently, the possible uniqueness of each person's fingerprints is somewhat of a straw argument in favor of fingerprinting as a forensic identification technique.

The attention placed on resolving the difficulties associated with latent prints virtually always being fragmented, degraded, "blurred, smudged, overlaid upon one another, incomplete, and distorted by foreign particles and dirt,"[486] takes on special meaning when the history of fingerprint evidence being allowed by the courts is understood.[487] It is known that in the early 1900s courts were induced to accept fingerprint evidence as legal proof establishing a person's identity, based on skewed probability estimates of their uniqueness.[488] It is noted in *Suspect Identities: A History of Fingerprinting and Criminal Identification* that all those estimates were based on "the likelihood of *whole* single fingerprints matching exactly in every particular. They completely overlooked the question relevant to forensic identification, which entails matching degraded fingerprint fragments."[489] Thus the vitally important unasked question that allowed fingerprint testimony to gain a foothold as scientific is: Can a partial latent fingerprint of one person,[490] which is significantly degraded in quality, be so similar to that of other people that it could be confused as originating from one of those other people?[491] The answer to that question is a resounding and unequivocal – *Yes!*[492]

IV – Conviction Based On Fingerprint Testimony Affirmed On Appeal In 1911

The possibility a latent print can mistakenly be linked to a suspect/defendant underscores there has been a lack of serious scrutiny of fingerprinting as an exact identification method from the very beginning of its use in courtrooms in the U.S. and other countries.[493]

The first criminal conviction in this country dependent on fingerprint evidence that was affirmed by an appeals court, was that of Thomas Jennings for the murder of Clarence Hiller by an intruder in his home on September

19, 1910.[494] There was no direct evidence that Jennings was the murderer, and the circumstantial evidence was so insubstantial that the prosecutor sought to link him to the scene of the crime by matching his fingerprints to impressions believed to have been left by the intruder in the *dried* paint of the victim's porch railing.[495] The prosecutor relied on the testimony of five fingerprint 'experts' who tried to emphasize that their testimony matching Jennings print to the impressions in the *dried* paint wasn't subject to doubt because it was based on certainty.[496] One of the examiners expressed this attitude by testifying: "I am positive. It is not my opinion."[497]

Yet it is reported that one fingerprint examiner "stated after examining the [fingerprint] photographs that the Chicago police had the wrong man."[498] That determination is consistent with the fact that none of the witnesses in the Hiller home saw the murderer's face.[499] They were only able to describe him as a "colored" man, and none positively identified Jennings at his trial.[500]

However for reasons unknown, Jennings' lawyer did not call the dissenting examiner as a witness to raise doubt about the certainty expressed by the prosecution's fingerprint "experts."[501] After finding Jennings guilty, several jurors were quoted in the *Chicago Examiner* as saying: "the finger-prints, and the finger-prints alone, convinced us that Jennings was the slayer of Hiller."[502]

In a display of judicial ignorance that continues to this day about the lack of solid science underlying the blind faith ascribed to the accuracy of fingerprint testimony, the judge in the Jennings trial ruled: "it is now an established fact that ... the lines upon one's fingers are different from that of the fingers of any other human being."[503]

In December 2011 the Illinois Supreme Court upheld Jennings' conviction by ruling the jury's reliance on "expert" testimony, and not their own observations about the fingerprint evidence, was proper because "the classification of finger-print impressions and their method of identification is a science requiring study."[504] That court's branding of a fingerprint evaluation as "science" opened the door to its admissibility by courts around the country as evidence that required interpretation by expert testimony.[505] So without any scientific proof being considered by the Illinois Supreme Court that fingerprints are in fact unique or that their identification involves an objectively duplicatable process, fingerprinting was christened as a 'science' that requires an expert interpreter.[506]

Thus was born "fingerprint voodoo." It is an idea so ingrained in the consciousness of the public, jurors and judges that since the earliest days of its use no jury in this country is known to have decided contrary to expert prosecution testimony tying a defendant's fingerprints to incriminating evidence.[507]

After his conviction was upheld, Thomas Jennings was hanged on February 16, 1912.[508] Doubts about his guilt persist to this day.[509]

V – When Are Fingerprint Samples Close Enough?

The number of points that must *be considered* to coincide between two samples for a fingerprint examiner to *feel comfortable* in testifying they match has long been a matter of contention.[510]

Influenced by French criminologist Alphonse Bertillon's demonstration in the early 1900s that two *different* fake prints "showed sixteen matching points of similarity," England adopted that as the minimum standard for admissibility of a fingerprint comparison as conclusive evidence.[511] Any fingerprint analysis that couldn't demonstrate a minimum of 16 matching points was automatically considered inconclusive.[512]

In contrast to the cautious approach of the British, no such minimum national standard was adopted in the United States.[513] Various law enforcement agencies did however, adopt informal guidelines ranging from 7 to 12 points of similarity, although matches are known to have been declared with as few as three concordant details.[514] Then in the 1940s the FBI abandoned any pretense that fingerprint analysis was an objective activity, when it adopted the policy that no minimum number of similar characteristics between a latent print and a suspect's fingerprint is required to declare a match.[515] So for over half-a-century defendants have been convicted in state and federal courts based on an FBI lab technician's testimony that one of the person's fingerprints matched a latent crime related print, based on nothing more than the examiner's subjective belief the prints were similar.[516] That is particularly significant considering United States District Court Judge Letts' observation in *U.S. vs. Parks* (1991), that fingerprint analysts' "...expertise is as fragile as any group I've ever heard hold themselves out as experts."[517]

An even more fundamental issue is that since it has never been scientifically determined that a person's fingerprints are unique, a false positive is possible when safeguards far more stringent than those in place in this country are enforced.[518] False positives are known to have occurred in England where a fingerprint examiner must successfully complete a multi-step training program that includes producing quality work during an on-the-job probation period before being allowed to testify in court after a minimum of three years experience: Even then they can only do so with the approval of the head of the fingerprint bureau *and* the chief of the police agency involved.[519] In contrast to those safeguards based on knowledge that examiner errors are a real problem, anyone in the U.S. can testify as a fingerprint expert who is permitted to do so by the trial judge.[520] That harkens back to the days when a person could acquire credentials by mail as an official finger-print examiner (FPE) by graduating from a correspondence course.[521]

VI – Fingerprint Fakery Has Been Known For A Century

The title of the 1924 book *Finger-Prints Can Be Forged*, sums up another danger to an innocent person made possible by a belief in the veracity of fingerprint evidence.[522] One of the authors, Albert Wehde, was a photographer and engraver who perfected a method of creating fake latent fingerprints.[523] Some examiners dismissed his prints as crude.[524] However some examiners were unable to distinguish between Wehde's fakes and real prints, so the critics may have been dismissive of his work to allay fears of jurists, jurors, defense lawyers, and the public at large about the danger posed by forgeries to the use of fingerprints as legal identifying evidence.[525] That is substantiated by the fact that fingerprint examiners were openly fearful of the danger investigators such as Wehde posed: After meeting during the 1927 national meeting of the International Association for Identification (IAI), the Ethics Committee issued a recommendation: "that every possible effort should be made to checkmate these activities insofar as they may prejudice the public against latent fingerprints found at the scene of crime as competent evidence in a criminal trial…"[526] They had all the more reason to be concerned because Wehde's book followed on the heels of reports concerning several other successful fingerprint forgery techniques.[527]

E. O. Brown

Former secret service agent, E.O. Brown developed a fingerprint forgery method that he demonstrated with dramatic flair in 1923.[528] Brown planted a fake print of the Berkeley, California police chief at the scene of a burglary.[529] The expertness of Brown's forgery technique was acknowledged, but the warning it sounded about the veracity of fingerprint evidence was ignored, because to not have done so could have undermined its use as an indentifying tool in criminal cases.[530]

Milton Carlson

E.O. Brown's demonstration was preceded in 1920 by the transferring of a fingerprint to a knife from a newspaper photograph by chirographer Milton Carlson.[531] Carlson's method allowed a person's fingerprint to be planted on an incriminating object if a photo of a person's print was available to a forger.[532] Based on his extensive knowledge of handwriting reproduction, Carlson noted *it is much easier to fake a person's fingerprint than it is to forge that same person's handwriting.*[533] He observed that unlike the subconsciousness involved in handwriting, fingerprints are mechanical in nature, so "... to complete a *perfect forgery* of a finger-print in the *exact form* is as easy to make as any steel ruler, surveyor's tape, or a wheel within a wheel."[534] Carlson also issued a challenge to expert fingerprint testimony that has gone unanswered to this day: "If it can be proved beyond a doubt that the finger-print in question is the impression made from the hand and by *contact* of the hand of the defendant, then finger-print testimony is of some value. If the expert on finger-prints cannot *prove* its genuineness or falsity, his testimony is of no value."[535]

Theodore Kytka

Seven years earlier, in 1913, handwriting expert Theodore Kytka of San Francisco, discovered a process of transferring a person's fingerprint from one object to another. To prove his process Kytka claimed he perfectly transferred the fingerprint of a police detective to an incandescent globe.[536]

Alphonse Bertillon

Prior to that, French criminologist Alphonse Bertillon faked "two *different* fingerprints which ostensibly showed sixteen matching points of similarity."[537]

Bertillon's opposition to the veracity of fingerprints as an identification system is often overlooked by the most reputable proponents of fingerprinting.[538] His criticism was particularly meaningful because he was an advocate of anthropometry: a system of identification that used measurements of human bones to identify a person.[539] Bertillon began developing his process that he called Bertillonage in the 1870s, and it was "the first modern system of criminal identification."[540] So he wasn't opposed to using the human body as an identification medium: His objection was that fingerprints are an unreliable identification method because they can be duplicated artificially, and there is no basis to believe they are naturally unique in an objectively measurable way.[541]

The Red Thumb Mark

Predating those multiple criticisms was the cautionary theme of *The Red Thumb Mark* – Richard Austin Freeman's well researched 1907 detective novel revolving around the theme that a perfect thumb print found in blood at the scene of a crime cannot in and of itself be considered to be substantive evidence implicating a suspect.[542] The book's principal character, Dr. Thorndyke, made the commonsense observation: "But there is no such thing as a single fact that 'affords evidence requiring no corroboration.' As well might one expect to make a syllogism with a single premise."[543]

Fingerprint Forgery Confirmed In 1925

The theory of fingerprint forgability was laid by numerous people in at least three countries during the 20th century's first three decades.[544] The reality of fingerprint forgery began in 1925 when "the FBI identified a forgery by a law enforcement officer."[545] Furthermore, at the IAI's 1929 national meeting – two years after the organization scornfully responded to Albert Wehde's warning of the ease of forging fingerprints, and four years after the FBI's exposure of fraudulent fingerprint evidence – three attendees respectively reported law enforcement fingerprint fabrications uncovered in Kansas, New Mexico and Minnesota.[546] The actual number of law enforcement fingerprint forgery schemes is unknown, but between 1930 and 1960 the FBI exposed an average of *one every two years* – 15 in 30 years – involving police agencies in 13 states across the country.[547]

New York State Police Crime Lab

Given the rich history of fingerprint forgery techniques publicly demonstrated numerous times in the first several decades of the 20th century, and the known fabrication of prints by crime lab personnel throughout the century, the revelation in 1992 that the New York State Police Crime Lab had been faking fingerprint evidence *for at least eight years* is not particularly surprising.[548] Fingerprint forgeries were involved in at least 40 cases, including homicide cases.[549]

Two of the five state police officers convicted of perjury, evidence tampering and official misconduct were latent fingerprint examiners certified by the IAI.[550]

The forgery techniques included lifting a print from an inked fingerprint card on file and transferring it to crime scene evidence, and photocopying an inked print and labeling it as a latent crime scene print.[551] The investigation after their scheme was uncovered indicated they began forging fingerprints, and continued doing so, because it was so easy to do and get away with.[552]

It was also learned during the investigation that the officers involved in the forgery ring were very careless at concealing what they did from anyone who would have cared to look.[553] The openness of what they were doing is mentioned in the official report to New York's Governor Cuomo: "This indifference, in itself, strongly suggests that the individuals fabricating evidence on a *routine basis* had no fear of discovery and, except with a noted exception, apparently took few steps to cover their tracks."[554] There is no known suspicion by their supervisors, judges, prosecutors, defense attorneys, or reporters of what they were doing on a regular basis: "In their confessions, the troopers themselves acknowledged that they chose to fabricate fingerprint evidence because they knew it would go unquestioned, because it was so thoroughly trusted."[555] It is also noteworthy that the crime lab forgers were investigated and prosecuted by a special prosecutor – not one of the prosecutors who had been benefiting from the print fraud scheme for nearly a decade.[556]

Considering the ease of forging fingerprints, the minimal or non-existent security measures preventing it, the obvious advantages it provides the prosecution, and increased computerization of fingerprint images, the New York State Police Crime Lab scandal may merely hint at the commonness of the practice across the country.[557]

That was somewhat confirmed by a study of reported cases of fingerprint falsification by fingerprint examiner Pat Wertheim, who believes such cases in the 20th century could number in the thousands.[558] Given the ease with which the NY State Police Crime Lab technicians openly fabricated fingerprints in *at least* 40 cases over eight years, with no effective oversight by supervisors, prosecutors, judges, other police agency or private examiners, defense attorneys, or curious news reporters, there is every reason to believe the actual number of such falsifications could be so high as to be unsettling to the most avid proponent of fingerprint evidence.[559] That suspicion is supported by the fact that the New York scandal was uncovered by the Central Intelligence Agency (CIA) during its questioning of "New York State Police investigator David Harding, who admitted forging fingerprints on numerous occasions."[560] That information was provided to the U.S. Department of Justice, which notified the New York State Police on May 26, 1992, that it had evidence of a fingerprint fabrication ring operating within the agency.[561]

Without the CIA's intervention it is unknown how much longer the forgery ring could have continued undetected, but there is no reason to think

Menace To The Innocent

it would not still be carrying on today. Which raises the disturbing question of how many crime lab fingerprint fabrication schemes are currently operating below the radar screen.[562]

Faking Fingerprints Is Easy And First Reported In 1913

It goes without saying that people outside of law enforcement, such as a master forger like Elmyr de Hory who had the skill to reproduce complex great works of art that were virtually indistinguishable from the originals, would find making a fake latent copy of a fingerprint a casual exercise.[563]

Not only is it known how easily an innocent person can be incriminated with a fake fingerprint, but as previously noted, a technique for transferring a person's fingerprint to an incriminating object the person did not handle was first reported in 1913,[564] and that was the theme of *The Red Thumb Mark* (1907) written six years prior to that.[565]

VII – Computerization Intensifies Questionability Of Fingerprint Evidence

With today's sophisticated computer equipment, creating authentic appearing fake fingerprints and transferring actual fingerprints to incriminate an innocent person is literally child's play for a skilled laboratory technician.[566] It is no more difficult than faking an incriminating photograph pawned off as original,[567] or prejudicial medical and police records indistinguishable from the "real thing."[568] However, just as a person doesn't have to be a photography expert to make convincing fake photographs with computer software, there is nothing barring a determined amateur with a scanner, off the shelf software, and a printer from digitally creating a convincing fake fingerprint that could be transferred to a potentially incriminating object or document, just as crime lab technicians are known to have done using conventional techniques since at least 1925 to frame a suspect.[569]

However the danger of computerization extends far beyond the generation of doctored or outright faked fingerprints by malevolently intentioned forensic technicians with access to a well-equipped laboratory, or a resolute lay person competently able to use readily available computer imaging programs.[570]

An additional danger lies in the unreliabilities inherent in the digitization of existing hard copy fingerprint cards, and/or the electronic scanning of fingerprint images.[571] As the national repository for electronic fingerprints, the FBI uses an Automatic Fingerprint Identification System (AFIS) to troll through the over 200 million fingerprint images in its database for a match with a scanned latent image.[572]

A major flaw in the AFIS system is that in spite of sounding precise, computerized matching systems have the same deficiency as the human examination process that doesn't require exactitude of compared print samples: Fingerprints are declared to "originate from the same source if they are "sufficiently" similar."[573] A central issue of both analysis techniques is what degree of similarity – not exactitude – is considered acceptable prior to declaring the scientific fiction that they match.[574] The probability of a mismatch occurring when a rolled ink print is manually compared with a fragmented and degraded latent print is multiplied to an unknown degree when an AFIS process is used to analyze a digitized sample print with a digitized latent print.[575] As inexplicable as it may seem at first glance, there are at least seven analysis distortion factors (ADF) that contribute to a comparison of digitized prints being no more, and in any given case, predictably less reliable than a human comparison of physical print samples.[576] Five of those ADF factors are:

- Sensory Distortion occurs when a finger is in contact with the optical surface during the electronic sensing/scanning process.[577] The distortion is caused by a combination of the pressure and angle of the finger on the scanning surface, and from the finger being three-dimensional, whereas the scanned image is two-dimensional.[578] Since different portions of its ridges are displaced "by different magnitudes and in different directions" on the hard surface (such as glass) it is pressed against, the elasticity of a fingertip's fleshy surface presents accuracy problems more serious for digitization of fingerprints, than the same type of distortion that occurs when they are manually rolled.[579]

- Irregular contact of a finger with the optical scanning surface results in an image distorted by omission.[580] Uneven contact is caused by such factors as "dryness of the skin, shallow/worn-out ridges (due to aging/genetics), skin disease, sweat, dirt, and humidity in the air..."[581] The inaccurate recording of a fingertip's details caused by irregular contact during the scanning process also occurs during the manual fingerprint rolling process.[582] The absence of full, even contact with a scanning surface, just like uneven pressure in applying the ink used to make a manual print impression, "results in "noisy," low-contrast images, which leads to either spurious or missing minutiae."[583]

- Non-reproducibility of a finger's features occurs when factors such as manual work, or an accident, permanently or semi-permanently causes alteration of its ridge structure.[584] It can also result in the recording of "spurious fingerprint features."[585] The phenomenon of non-reproducibility is not unique to scanned fingerprints, since it also affects the identifiability value of manual inked prints.

• Measurement errors attributable to an imperfect image-feature extraction algorithm is a problem unique to scanned fingerprints.[586] This can result in inaccurate "location and orientation estimates" of a fingerprint's structure and patterns.[587]

• Sensory noise is added to a fingerprint image by the scanning process.[588] One way this is manifested is the geometric distortion of the fingerprint due to imperfect imaging conditions.[589] Another way is residue can remain on the sensing surface from a previous scan.[590] In a manner of speaking, sensory noise is not necessarily limited to scanned images, because residue can also exist on an inked print surface, or it can be applied to the surface after being picked up from the ink source.

Singly or in combination, those five factors undermine the reliability of a match between a digitized fingerprint and a digitized latent print, and to the degree applicable, a rolled print's comparison to a latent print. However, as prejudicial as those factors are to the reliable matching of electronically scanned fingerprint samples from different sources, they pale in significance next to the likelihood of error caused by the *sixth ADF*: computerized enhancement of a scanned print, that can be compared with a likewise enhanced latent print.[591]

Prior to the development of digital enhancement techniques, a latent print was subjected to a two-step filtering process: Fingerprint examiners *magnified* a latent print *made visible* by a conventional process.[592] Although neither of those processes involves the deliberate substitution of a fingerprint's known features for ones that have more clarity, that is precisely what is done by computerized fingerprint enhancement techniques designed to alter a latent print sample to increase the degree and/or frequency of its features.[593]

Enhancement techniques were developed to reconstruct a fragmented, partial and/or degraded print[594] by extrapolating what it is projected to have looked like if the full print had been obtained in a controlled environment.[595] This is done by first dividing a fingerprint image into small blocks: The shape and frequency of ridges and valleys, and their orientation to each other is then estimated from the available information, and filled into the image as if it existed in the original.[596] When the missing portions of a fragmented or degraded latent print are restored, it is ascribed an evidentiary value to the degree it can successfully be matched with a known fingerprint sample.[597]

The enhancement process is not limited to latent fingerprints.[598] It is recognized that 5% or more of people – 1 out of 20 – have fingerprints with features too obscure to reliably be analyzed without enhancement by "automatic image processing methods."[599] However significant that percentage may seem, the authors of *Advances in Fingerprint Technology* think it may underestimate the actual percentage of people whose fingerprints when obtained under ideal conditions, are too degraded to be analyzed without computer software enhancement.[600] They write: "We suspect this fraction is even higher in reality when the target population consists of (1) older people, (2) people who suffer routine finger injuries in their occupation, (3) people living in dry weather conditions or having skin problems, and (4) people who have poor fingerprints due to genetic attributes."[601]

Whether or not a person has fingerprints with indistinct features contributing to a low scanability quotient, low-quality optical scanners are an additional contributor to the creation of indistinct digitized prints requiring a high degree of enhancement.[602]

Yet computer generated restoration techniques undermine the claim fingerprints are unique: If a fingerprint is comprised of non-repeated patterns, then it is not possible for its missing "unique" sections to accurately be reconstructed.[603] At best a restored print can only represent a *guesstimation* of what it would be expected to look like based on *known* fingerprint patterns, because a unique but incomplete fingerprint is unknowable without its missing features being available for analysis.[604] Thus the argument of fingerprint uniqueness precludes restoration of a fragmented latent sample, and advocacy of reconstruction techniques implies conceding fingerprint commonality.

The quality of a digitized fingerprint's details is further affected by the *seventh electronic analysis distortion factor*: the use of a compression scheme to reduce an image's file size. [605] Experienced computer users are aware of the advantages of compressing file sizes to reduce storage space, for faster emailing of a file as an attachment, and to shorten download times. To make such tasks easier to perform, many popular computer programs – including the world's largest selling operating system – incorporate use of the ZIP compression scheme as a command option.[606] ZIP uses what is known as a *lossless* compression method, because no data is omitted to

Menace To The Innocent

reduce the file size, and hence the file is perfectly recreated when it is decompressed.[607]

Compression is integrally related to fingerprint digitization, because the the FBI standard for a scanned print is 500 DPI with 8 bits of gray scale.[608] That results in a file size for each print of about 10 MB.[609] The over 200 million digitized fingerprints in the FBI's database, would thus require more than 2,000 terabytes of storage space for those uncompressed images.[610] Furthermore, since the FBI's fingerprint database is growing by 30,000 to 50,000 images per day – an additional 300 to 500 gigabytes of storage space per day would be required for uncompressed prints.[611]

However a problem with compressing an image file is the maximum it can be compressed without a loss of details by a lossless method, is about 2 to 1.[612] Although that would result in a reduction in storage space from roughly 2,000 to 1,000 terabytes for the FBI's fingerprint database, the FBI did not consider that to be a satisfactory rate of compression to reduce storage requirements, and facilitate rapid AFIS comparisons.[613]

To achieve a higher compression ratio the FBI decided to omit details from a print's image by what is known as a *lossy* compression method.[614] Lossy compression schemes achieve high compression ratios of image files by progressing from omitting the smallest and highest frequency details, to those that are ever larger and less frequent.[615] What that means is loosy compression schemes work by systematically distorting an image from its original form, and the higher the rate of compression the greater the degree of the distortion.[616]

A compression ratio of 12.9 was settled on by the FBI.[617] The JPEG compression scheme – widely used to compress images on World Wide Webpages – was tried, but it was considered unsuitable due to the distortion caused by omitting too many critical details.[618]

Another lossy method – WSQ[619] – creates file sizes virtually the same size as JPEG at a 12.9 compression ratio.[620] Nevertheless, the FBI chose it as their standard.[621] However, WSQ, like JPEG, compresses by omitting details, and anyone who has compressed an image to a significant degree knows how rapidly the sharpness of its details are degraded. That is because lossy compression schemes work by first omitting the "little" details that make an image sharp – while retaining its overall shape.[622] Yet given that a cross-section of any person's fingerprint can be expected to match that of any number of other people,[623] the only reasonable hope to ensure distinguishment between them is preservation of every image detail which the WSQ method – just like JPEG – systematically omits.[624] The distorting effect of lossy compression on a scanned fingerprint image is compounded by the application of suspect enhancement techniques.[625] The compression of a fingerprint's details is the *third* generation of digital distortion that follows the *second* generation distortion caused by computerized enhancement, that follows the *first* generation distortion caused by the scanning process.[626]

The consequences of law enforcement's use of a lossy compression scheme is an increased likelihood of an honest mismatch occurring by a conscientious lab technician. However it also makes it that much easier for a suspect to conveniently be confirmed as the culprit by a prosecution biased examiner.[627]

Another side-effect of using electronic analysis techniques is they can induce mental laziness by the substitution of computerized analysis for the thinking of a lab technician.[628]

A significant negative effect on the reliability of fingerprint testimony can result when the mental stultification associated with a reliance on computerized fingerprint analysis techniques is affected by fingerprint distorting factors such as: indistinct fingerprint features; low-quality scanning; digitized fingerprint enhancement; and compression of fingerprint images. That negative effect reinforces the voodoo like nature of fingerprint examinations.

VIII – The *Mitchell* Case (1999)

One way fingerprint analysis' lack of soundness as an identification process would be expected to be reflected, is in a predictable error rate.[629] That became an issue during a pre-trial *Daubert* challenge[630] to the scientific basis of fingerprint testimony in *U.S. v. Mitchell*, CR No. 96-407 (E.D. PA).[631]

In the 1993 case of *Daubert v. Merrell Dow Pharmaceuticals*, the Supreme Court established standards for what constitutes expert scientific testimony under *Federal Rule of Evidence* 702.[632]

To substantiate the expertness of fingerprint testimony, the FBI sought to aid federal prosecutors in *Mitchell* by seeking to provide proof that fingerprint analysis had a very low or non-existent error rate.[633] To do this, the FBI crime lab sent the defendant's ten-print (all ten fingers) card and latent prints lifted from the robbers' abandoned getaway car, to law enforcement crime labs around the country for comparison.[634] The response was unexpected by the FBI: 9 of the 34 crime labs responding – 26.5% –

were unable to match one or more of the defendant's prints to those found in the alleged getaway car.[635] The negative response rate was particularly significant for two reasons: the test was not blind; and the lab *only* had the yes/no option of deciding whether the prints matched.[636]

The lack of crime lab unanimity in responding to the *Mitchell* test was unacceptable to the FBI.[637] The agency attributed the responses of the nine non-conforming labs to factors such as the examiners: "just screwed up";[638] were "inexperienced";[639] devoted "insufficient time" to making the examinations;[640] and "were probably tired" if the examinations were conducted "late in the day."[641] Thus the FBI attempted to excuse the high rate of dissimilar results in the *Mitchell* test to a combination of incompetence, sloppiness, indifferent work attitudes, lack of qualifications, fatigue, and conscientious error by the participating crime lab technicians.[642] It was a telling admission for the FBI to explain that slovenly conduct, bad attitudes, and a lack of skill affected the analysis of a suspect's prints and latent crime scene prints, when it was known by the crime lab examiners that the results would be submitted by prosecutors in a federal case to support the scientific exactitude of fingerprint examinations and testimony.[643]

The FBI needed to control the damage caused by the *Mitchell* test results to their claim that fingerprint identification has the exactitude of a scientific discipline. Consequently, the FBI made two significant changes to the original photos of the fingerprint samples, before sending them back to the nine labs that provided contrary results for re-examination.[644] Those two changes were: the photos were greatly enlarged; and, "red dots" marked the "Galton points" where the FBI believed Mitchell's prints matched those found in the getaway vehicle.[645] The FBI also emphasized the re-examination results would be incorporated into the prosecution's response to the defense's fingerprint challenge in the *Mitchell* case.[646]

The FBI defended its coaching of the technicians in the nine crime labs, and its not so subtle pressure on them to produce a match of the prints on the second go-around, by comparing them to FBI fingerprint trainees who need hands-on coaching to produce a result identical to their supervisor.[647] When those nine labs dutifully complied by matching the suspect's prints with the latent crime prints, the prosecution was able to submit documentation showing a 100% compliance rate, and the judge in *Mitchell* subsequently ruled in the government's favor.[648]

The FBI's coaching and pressure tactics to achieve unanimity in the *Mitchell* test takes on special meaning when it is considered the original

26.5% rate of crime lab non-unanimity was consistent with the error-prone results of zero-blind fingerprint examiner proficiency tests from 1983 to 2001.[649] Which is put in additional perspective by the fact that although fingerprint evidence has been admitted in U.S. courts since 1910, there has never been a scientifically sound double-blind test to ascertain the skill level of fingerprint examiners.[650] The high error rate in tests tilted towards generating low error rates, are an indicator that for nearly a century the wool has been pulled over the eyes of state and federal judges, jurors, defense lawyers, as well as the public, about the supposed infallibility of fingerprint analysis.[651] This mythology is so ingrained that an FBI lab supervisor testified in the *Mitchell* case that fingerprint methodology is so perfect that the rate of error in his examinations is *"zero."*[652]

IX - The *Plaza* Case (2002)

In January 2002 many people first became aware that claims of fingerprint matching exactitude might not be accurate, when the national press reported on a *Daubert* challenge to the admissibility of expert fingerprint testimony in *U.S. v. Plaza.*[653] The challenge was similar to that in the *Mitchell* case, except that U.S. District Court Judge Pollak initially granted a defense motion to bar testimony that "in the opinion of the witness, a particular latent print is – or is not – the print of a particular person."[654] Judge Pollak reversed his decision two months later, after the prosecution requested reconsideration of his order.[655] In his new order he allowed testimony of a fingerprint examiner as to whether a match did or did not exist between the defendant and a latent crime scene print.[656] Judge Pollak reversed himself based on the FBI's reliance on an examination process that excluded a minimum point standard for a match to be declared, which was "essentially indistinguishable" from the standard adopted by Scotland Yard in 2001.[657]

X – The *Parks* Case (1991)

Considering the publicity given to *Plaza* and similar cases, it is noteworthy that a successful challenge to fingerprint evidence came in a federal bank robbery case in 1991 – two years before *Daubert*.[658] In that California case, *U.S. v. Parks*, U.S. District Court Judge Spencer Letts excluded testimony by the prosecution's experts based on the lack of standards in determining a match between two fingerprints.[659] During Judge Letts' questioning, one of the prosecution's experts approvingly testified about the lack of a minimum standard for the number of points that must match before a suspect's print and a latent print are declared to be the same:

The number of matching points varies widely between different cases, different prints in the same case, and different police agencies.[660] Judge Letts observed there is what he referred to as the "sliding scale"[661] of points considered necessary to match for an identification to be made: "You don't have any standard. As far as I can tell, you have no standard. It's just an *ipse dixit*. "This is unique, this is very unusual." "How do you know it's unusual?" "Because I never saw it before." Where is the standard, where is the study, where is the statistical base that been studied? The FBI has zillions of these things, where is a study of the entire computer bank?"[662] Judge Letts further noted in regard to fingerprint analysis: "I will say, based on what I've heard today, the expertise is as fragile as [for] any group I've ever heard hold themselves out as experts."[663]

The situation is exactly the same today as in 1991, except that since then no judge, even after applying the *Daubert* test, has made a final ruling related to the admissibility of fingerprint evidence consistent with its lack of a scientific basis that from his reaction, appeared to be virtually self-evident to Judge Letts.[664]

XI – Erroneous Fingerprint Identifications Are A Known Problem

The malleability of a fingerprint examiner's subjective conclusion has led to erroneous fingerprint identifications in a number of known, and an untold number of unknown cases. The following seven cases hint at the scope of the problem:

John Stopelli

John Stoppelli was convicted in the late 1940s of being involved in a heroin ring in Oakland, California based on a fingerprint identification, and sentenced to six years in prison.[665] After losing his appeals, Stoppelli appealed to President Truman for a pardon based on subsequent fingerprint examinations that disclosed the errors in the initial analysis; affidavits from four of the drug ring's members that he wasn't involved; and incontrovertible proof that he was 3,000 miles away in New York City, where he lived, at the time of the crime.[666] President Truman granted John Stoppelli's pardon request, and he was released after serving two years in prison.[667]

Bruce Basden

When questioned by police after trying "to pawn an expensive ring" that had belonged to a woman who was murdered in Fayetteville, North Carolina in June 1985, Joseph Vestal claimed it was a gift from an acquaintance,

Bruce Basden.[668] Vestal further claimed that Basden, a U.S. Army Sergeant stationed near Fayetteville at the time of the crime, boasted "it was part of the loot from his robbery and murder" of the woman and her husband.[669] A fingerprint found at the crime scene was identified by state crime lab technician John Trogdon as Basden's, and he was extradited from Texas where he was then stationed.[670] In response to the request of Basden's lawyer to have the fingerprint evidence reappraised, Trogdon enlarged the latent print photos to enhance their details.[671] That led to his discovery of dissimilarities with Basden's prints that he had previously overlooked and caused him to change his mind that Basden was 'the guy.'"[672] The prosecution dropped the charges against Bruce Basden and he was released after languishing in jail for 13 months.[673]

Trogdon's justification for his error was reported in *The Scientific Sleuthing Newsletter*: "The points of similarity … were discernible at all times, he said; it was the points of dissimilarity that did not spring into view until the enlargements were made."[674] It was found after a limited review of other cases Trogdon was involved in, that he had identified "three fingerprints [that] did not belong to three defendants in three separate cases."[675] It is unknown how many more defendants would have been exculpated instead of inculpated by Trogdon, if he had enlarged the suspect's fingerprints in every case.

Michael Cooper

In May 1986 three different examiners cross-verified that latent prints collected from two different rape scenes in Tucson, Arizona originated from Michael Cooper.[676] The only problem with tagging Cooper as the "prime time rapist" is it was subsequently proven he was innocent, and that the crime scene prints weren't his.[677] The *Arizona Republic* reported on December 3 1986 that the three crime lab technicians involved in Michael Cooper's misidentification were disciplined with a demotion and/or a suspension.[678]

Roger Caldwell

Roger Caldwell's 1978 conviction of a 1977 double murder in Duluth, Minnesota hinged on a Minnesota state fingerprint examiner's testimony that Caldwell's right thumb print, "was identical"[679] to a photo of a print found on the back of an envelope containing a gold coin belonging to one of the victims.[680] Yet a year later his co-defendant, who was his wife, "was acquitted of all charges of aiding and abetting and conspiracy to commit murder."[681] Among the testimony during the trial of Caldwell's wife in 1979, was that examination of the *negative* of the envelope thumb print photo excluded him as its source.[682] However neither the negative of the thumb print photo,[683] nor the statement of a credible alibi witness establishing Caldwell was at a Golden, Colorado hotel – *811 air miles* distant from Duluth – four hours before the 2am murders,[684] was provided to him by the prosecution prior to his trial.[685] In reversing Roger Caldwell's conviction after he had served five years of his life sentence, the Minnesota Supreme Court stated of the prosecution's fingerprint evidence: "The fingerprint expert's testimony was damning – and it was false."[686]

Richard "Riky" Jackson

In 1998 Richard "Riky" Jackson was convicted of the September 1997 murder in Philadelphia of an acquittance. His conviction was based on the testimony of three prosecution fingerprint experts that the bloody fingerprints left on a box fan at the crime scene matched Jackson's on at least eight points.[687] Yet their certainty of a match was disputed by two experts who testified for the defense that Jackson's prints and those of the killer couldn't be more dissimilar: One of those experts, who retired the year before from a long career in the FBI's Latent Fingerprint Section, said the prints "weren't even close."[688]

After Jackson's conviction and sentencing to life in prison, his lawyer asked the International Association of Identifiers (IAF) to examine the print samples the jury relied on.[689] A panel of IAF examiners were unable to match Jackson's prints to those found at the murder scene.[690] When questioned by the IAF, the prosecution's star forensic witness acknowledged he could have been mistaken.[691]

In response to Jackson's post-conviction motions disclosing the fingerprint discrepancies, the prosecution had the prints reanalyzed: When that produced a negative match, the FBI crime lab was consulted, and it too excluded Jackson as the origin of the crime scene prints.[692] With the sole physical evidence against Jackson discredited, the prosecutor moved to vacate his conviction.[693] "Riky" Jackson was released in December 1999, after more than two years of wrongful imprisonment.[694]

Stephan Cowans

Stephan Cowans was convicted in 1998 of non-fatally wounding a Boston policeman with his pistol during a struggle.[695] The jury relied on expert prosecution testimony that Cowans' left thumb print matched a latent print found on a glass mug used by the assailant.[696]

In response to Cowans' post-conviction challenge to his conviction, in May 2003 a court ordered DNA testing to be performed on the mug and several items of clothing worn by the assailant.[697] When Cowans was excluded as the source of the DNA on those items, the prosecutor responded in January 2004 by requesting that a sweatshirt worn by the assailant also be subjected to DNA testing.[698] When that test also excluded Cowans, the prosecutor had the fingerprint evidence the jury relied on re-analyzed.[699] The result was negative.[700] So contrary to the expert testimony at Cowans' trial, the crime scene print was dissimilar to his thumb print.[701] Faced with proof positive by their own experts of Stephan Cowans' innocence, the prosecution dropped its opposition to vacating his conviction.[702] He was released on February 3, 2004 after serving 6-1/2 years of a 30-45 year prison sentence.[703]

Stephen Cowans' case is another illustration of the power of prosecution expert testimony to overwhelm other seemingly solid evidence pointing to a person's innocence, or at the very least, casting a reasonable doubt on that person's guilt: The family in the home of the wounded policeman had spent considerable time with the assailant, but they did not identify Cowans as that person during a lineup.[704]

Brandon Mayfield

Brandon Mayfield, a Portland, Oregon attorney, was arrested on May 6, 2004 for his suspected involvement in the March 11, 2004 bombing of four commuter trains in Madrid, Spain that killed 191 people.[705]

The Spanish National Police (SNP) detected fingerprints on a plastic bag containing detonators found in an abandoned van near the departure point of three of the trains.[706] After examining the prints at the request of the SNP, the FBI arrested Mayfield on May 6, 2004 as a material witness.[707] In the arrest warrant affidavit alleging Mayfield's suspected involvement in international terrorism, an FBI examiner swore that Mayfield's left index fingerprint matched one found on the plastic bag "in excess of 15 points of identification."[708] The affidavit also stated that Mayfield's fingerprint identification was verified by an FBI fingerprint supervisor and a retired FBI fingerprint examiner with 30 years experience who was on contract with the FBI's lab,[709] and "… the FBI lab stands by their conclusion of a 100 percent positive identification."[710] In addition, on May 19, a fingerprint examiner hired two days earlier to provide the federal judge assigned Mayfield's case with an independent expert opinion about the FBI's identification of Mayfield's fingerprint, testified by telephone that it did indeed match the

64   Menace To The Innocent

print found on the plastic bag.[711]

At his first court hearing, Mayfield, who had an expired passport and had never been to Spain, told the federal judge: "That's not my fingerprint, your honor."[712] He was right. On May 20, 2004, the SNP announced that two prints on the bag had been linked to an Algerian with a police record and a Spanish residency permit.[713] The next day, May 21, Mayfield was conditionally released, and on May 24 the warrant was dismissed.[714]

After Mayfield's release, the FBI claimed Mayfield's misidentification was due to a "substandard" fingerprint image.[715] However that claim was contradicted by former Scotland Yard fingerprint examiner Allan Bayle, an internationally recognized expert with more than a quarter century of experience who was retained by Mayfield's public defenders.[716] Bayle determined the clarity of the Madrid fingerprint photo was good, and that it was so dissimilar from Mayfield's print that they shouldn't have been declared a match by a competent examiner.[717] He said of the FBI's analysis: "It's flawed on all levels," and he described it as "horrendous."[718] That analysis was consistent with the SNP's, which reported to the FBI on April 13 – 23 days before Mayfield's arrest – that their comparison of his fingerprint with the one on the plastic bag was "conclusively negative."[719]

Brandon Mayfield's fingerprint misidentification is significant because four fingerprint examiners – two employed by the FBI and two independent experts (one under contract to the FBI) – all confirmed that his left index finger matched the print on the plastic bag when it didn't.[720] Mayfield's federal defender, Steven Wax, observed that his release was due to the SNP's crime lab's exclusion of him, and its public announcement that another suspect had been identified: "But for the unusual circumstance of another national police agency conducting its own independent investigation, Mayfield would still be incarcerated."[721] Mayfield's other attorney, federal defender Chris Schatz, openly wondered how many people didn't have a White Knight to save them from a police crime lab's false fingerprint identification: "Who knows how many people are sitting in state and federal prisons that have just never come to light because there is no independent agency like the Spanish National Police."[722]

Mayfield filed a lawsuit against the federal government.in federal court in June 2004. Mayfield hired well-known attorney Gerry Spence as his lead lawyer. Spence said he agreed to become involved because: "Our basic rights under the Constitution are in jeopardy, and that's what this is about."[723]

In Novemer 2004 a report by a seven-member panel of international experts was released that analyzed the FBI crime labs performance in the Mayfield case.[724] The panel, assembled by the FBI in June 2004, determined the crime lab's culture of deference to an examiner's initial determination caused Mayfield's erroroneous identification to be unchallenged even if the error was detected by subsequent examinations: "To disagree was not an expected response."[725] The panel also found that when the SNP disagreed that the print was Mayfield's, the FBI "entered into a defensive posture" instead of reevaluating its identificaton of him.[726] Furthermore, the panel found that the FBI's contention that the error was caused by a "substandard" print image was without merit.[727] The panel did not explore the possibility that the FBI's examiners were predisposed to identify the print as Mayfield's by overt or subtle suggestions telegraphed by others within the FBI.[728]

XII – Erroneous Fingerprint Identifications Also Occur In England

The total inadequacy of safeguards protecting the innocent in this country from erroneous fingerprint matches, whether they are accomplished manually, digitally, or by a combination of the two, is emphasized by the fact that until 2001 the United Kingdom applied a more stringent standard of proof than in the U.S. Yet, that didn't prevent the U.K. from having high profile cases of an innocent person's conviction being based on erroneous expert fingerprint testimony.[729]

One such case is that in 1982 the IRA took credit for a bombing that killed four policeman.[730] Five years later Danny McNamee was convicted of making the bomb based on expert testimony his fingerprints were "the same original master" as those found on a piece of tape and a battery at a bomb assembly location, and that they also matched a thumb print found amidst the bomb debris.[731] He was sentenced to life in prison.[732] In December 1998, after 11 years of imprisonment, the U.K.'s Court of Appeals quashed McNamee's conviction because it was proven the fingerprint evidence was wholly unreliable.[733] Of fourteen fingerprint examiners consulted during the appeal, the most points of similarity any could find between Danny McNamee's and the crime prints was 11, far less than the 16 matching point standard then required under British law, and two examiners said the poor quality of the latent prints precluded them from being used for identification.[734]

McNamee's exoneration is not an isolated incident: A year prior to his release Scotland Yard acknowledged that two examiners, *whose work was triple-checked*, had erroneously matched an innocent suspect's prints to

Menace To The Innocent

those of the person who burglarized writer Miriam Stoppard's home.[735]

Whatever its deficiencies, the British standard requiring 16 matching points provided a measure of protection from blatantly obvious mismatches between latent crime scene prints and those of a suspect.[736] However on June 11, 2001,[737] 81 years after its adoption, that standard was abandoned when Britain adopted the FBI's policy of not requiring any "set numerical standard to be satisfied before experts make a decision that a mark or impression left at a crime scene and a fingerprint were made by the same person."[738] It is sobering to think that Danny McNamee could have died in prison if his appeal had been considered only 30 months later than it was.[739]

McNamee's case particularly highlights the danger fingerprint evidence poses to the innocent, because his conviction hinged on that evidence, and there was no apparent technician subterfuge involved in applying the 16-point standard.[740] It merely takes a jury's reliance on the testimony of the prosecution's expert witness to seal the conviction of a defendant, particularly if the person can be made out to appear to be guilty for reasons unrelated to his or her fingerprints.[741] This danger is compounded by the mendacious, or even frivolous intentions of crime lab personnel and/or prosecutors.[742] Thus falsification of fingerprint evidence is a very convenient and effective way to ensure an innocent person's frame-up.[743] Particularly considering that its sacrosanct status as incontrovertible evidence among judges, jurors, prosecutors, and even defense lawyers, pressures an untold number of innocent people to falsely plead guilty when faced with a prosecution expert's testimony that the person's fingerprint matches a latent crime related print.[744]

In spite of its suspect foundation and the nefarious purposes it can be used for, fingerprint evidence is relied on so heavily as a law enforcement identification tool, that it is not expected to be replaced in importance by DNA as an evidentiary source for prosecutors.[745]

Chapter 7

DNA Probability Estimates Elevated By Smoke And Mirrors To Certainty

DNA evidence has been adopted as a scientific darling of prosecutors for the same reason as fingerprints: judges, jurors, reporters, and the general public are dazzled by its alleged infallibility.[746] It has achieved sacrosanct status because it is widely believed to be a method of irrefutably identifying the person from which a DNA sample originated.[747] FBI Director William Sessions expressed that belief in a speech to the National Press Club: "Probably the most exciting, as I view it, of the new techniques emerging for the criminal investigator is the DNA identification technology. Through a genetic pattern-matching process, criminals can now be *identified positively* by comparing evidence from a crime scene – that is, blood, body fluids, or sometimes a single hair – with that of a suspect."[748]

Yet contrary to the popular belief expressed by then FBI Director Sessions, DNA evidence does not and cannot establish the identity of a crime's perpetrator.[749] *At most* it can only identify a range of probability that a DNA sample related to a crime is consistent with a particular person's DNA profile.[750] The authors of *Tainting Evidence*, explain this in the following way:

> "In fact, DNA typing or profiling is not a genetic fingerprint, and to portray it as such is scientific fraud. ... DNA profiles are not [unique]. DNA typing produces what is known as a random probability match – sometimes as high as one in several million or even billion, sometimes as low as one in dozens. As such it represents the probability of a match between a sample left at the crime scene and a suspect, not a definitive match itself. To pretend, as some proponents seem to, that DNA profiling, this sci-crime wizardry is infallible is a gross distortion of the truth."[751]

Professor Dan L. Burk went further in decrying proponents of DNA's ability to make an absolute identification: "Bald statements or broad hints that DNA testing is infallible ... are not only irresponsible, they border on scientific fraud."[752] Such firm sentiments are rooted in the reality that DNA identification involves statistical probabilities – a form of mathematical guesstimating – not certainty.[753] So contrary to the mystical like reverence accorded the typing of human DNA that has continued to grow since it was developed in the 1980s, *at best* it can only narrow the possible number people a sample originated from, while excluding those it clearly did not come from.[754]

As with other testing processes, personal proclivities – what can be called the subjective element – can affect the final analysis of a test result.[755] Those are compounded by possible contamination during the collection, transportation, storage, and testing of DNA evidence.[756] This problem is so endemic that Washington State Patrol Lab documents reveal that in at least 23 serious felony cases, lab technicians contaminated DNA samples or made other errors in the handling of DNA evidence.[757] In eight of those cases the crime related DNA evidence was tainted by the technician's own DNA.[758] The errors in six cases included misreading test results and disposing of evidence swabs, three cases involved cross-contamination with DNA from an unrelated case, and the contamination source was undetermined in five cases.[759] Since the lab is only detecting errors in cases where the evidence may be contested, the actual number of errors is likely to be much higher, since nationwide 96% of state court convictions are by a guilty plea that limits critical examination of a crime lab's handling and processing of the prosecution's physical evidence.[760]

DNA's limitation as an identifying medium was recognized when it was first introduced as *confirmation* evidence, after the likelihood of an accused's guilt had been established by *other* evidence.[761]

The National Association of Criminal Defense Lawyers (NACDL) has warned that in spite of the subjective nature of DNA analysis: "the FBI wants to be able to state to a *"scientific certainty"* that two DNA samples match. This is an assertion no other forensic DNA laboratory would dare make – because there are no certainties in science, only probabilities."[762]

Thus a prosecutor's claim that a DNA "match" between a suspect and crime-related evidence proves a defendant is guilty beyond a reasonable doubt makes for a good sound bite, and might help convince a judge and jury to convict the person, but it isn't based on science: Yet DNA evidence alone is now considered legally sufficient to sustain a conviction in the absence of *any* corroborating evidence.[763]

In essence, DNA testing is a sophisticated form of blood or tissue analysis. A person can be excluded as a suspect if their blood type doesn't match that found at a crime scene. Similarly, if a person's blood type matches that at a crime scene, it doesn't mean she or he did it. The person is simply in the pool of people with that blood type who *might* be guilty. The same is true of DNA testing: along with blood typing it is a form of circumstantial – not conclusive – evidence.[764]

However DNA is a more subjective process than blood typing, because it involves a much less clearcut evaluation process, that is similar in principle to that involving fingerprints.[765] DNA analysis involves subjectively matching "bands" between samples, while fingerprint analysis involves subjectively matching "points" between samples.[766]

A defendant is linked to crime related DNA sample by what is known as a "random match probability."[767] That is an estimate of a *coincidental* match between two people with similar DNA profiles.[768] That estimate is expressed as a number – such as a million to one or a hundred million to one, etc. – that is intended to portray the likelihood the DNA evidence associated with the crime originated from the defendant.[769] Since it always sounds impressive, the news media is quick to report a crime lab technician's testimony about whatever that probability is guesstimated to be. This figure is always far in excess of the 99 to 1 standard of certainty the Supreme Court inferred in *Schlup v. Delo* (1995) is necessary to support a conviction.[770] So once this testimony is given, the defense bears a heavy burden to overcome.[771]

However the "random match probability" is only half the story of whether two samples have a common source.[772] Although invariably overlooked by observers, the other half is the "false positive probability."[773] The two probabilities starkly contrast with one another because the latter undermines the degree of certainty assigned to DNA evidence by the former.

Menace To The Innocent

Chapter 8

False Positives – DNA Testings Dark Side

A false positive occurs when a match is reported between a base sample and a test sample that actually have dissimilar DNA profiles.[774] An example is when the semen collected from a rape victim is used as the DNA base of comparison with a DNA sample (such as saliva or blood) from a suspect, to falsely implicate that person.[775]

That is exactly what happened to Josiah Sutton.[776] During his trial, a Houston's Police Crime Lab technician testified DNA evidence established he was "definitely" one of two men who raped a woman after a car jacking.[777] Sutton was subsequently convicted and sentenced to 25 years in prison.[778]

In the fall of 2002 Houston television station KHOU began an investigation into practices of the Houston Police Department Crime Laboratory.[779] The station used Texas' public records act to obtain laboratory files.[780] Needing an expert to analyze the documents, the station contacted UC Irvine Law Professor William C. Thompson after being informed he "is the best person in the country with regard to finding problems in forensic DNA lab work."[781] While reviewing the documents Professor Thompson found many critical irregularities concerning the operation of the HPD Crime Lab.[782] When reported by KHOU, those finding led to the DNA lab being shut down in December 2002, pending a complete review of its procedures and personnel.[783]

In January 2003 Professor Thompson was provided with Josiah Sutton's file.[784] After reviewing it he concluded that contrary to the technician's trial testimony, the DNA test result he relied on only established a 1 in 8 (12-1/2%) probability that Sutton was one of the rapists.[785] That is nowhere near the 99% certainty the Supreme Court inferred in *Schlup v Delo* (1995) is necessary to support a conviction.[786] Even more damaging for Houston's crime lab, is that further analysis of the test report led Professor Thompson to conclude it actually excluded Josiah Sutton.[787]

Retesting of the DNA sample by a *private laboratory* confirmed Professor Thompson's assessment.[788] Excluded as being one of the attackers, Josiah Sutton was released from prison in March 2003, after four years of wrongful imprisonment.[789] It is noteworthy that Sutton's case didn't happen in the early days of DNA testing when the excuse could have been made that chinks in the process needed to be worked out: It happened after a belief in the infallibility of DNA testimony had become widespread throughout the legal fraternity and society in general.[790]

Gilbert Alejandro was another victim of erroneous expert DNA testimony. Accused of sexual assault, during his 1990 trial the victim identified him as her assailant (in spite of testifying she had a pillow over her head during the assault), and the prosecution's expert witness, Fred Zain: "testified that a DNA test of Alejandro's sample matched DNA found on the victim's clothing "and could only have originated from him [Alejandro]."[791] Alejandro was sentenced to 12 years in prison.

When Alejandro's lawyers later learned that Zain that lied about his credentials and he had given false testimony in other cases, they filed a writ of habeas corpus requesting a new trial. During a hearing on July 26, 1994, the jury foreman and one juror testified they relied solely on Zain's testimony to convict Alejandro, and without it they would have voted to acquit him on the basis of reasonable doubt. A DNA analyst testified that results from at least one other DNA test excluded Alejandro, and the test on which Zain based his testimony was inconclusive, and did not provide a basis to associate Alejandro with the crime scene evidence. Alejandro's conviction was set-aside and in September 1994 the prosecution dropped the charge. He had been wrongly imprisoned for four years.[792] In June 1995 Alejandro was awarded $250,000 from a lawsuit he filed against Bexar County, Texas.

Josiah Sutton's case also illustrates how the reality of false positives undermines the misguided belief that testimony related to a random probability match between two DNA samples can be taken at face value.[793] It also emphasizes that DNA has only achieved the status of being considered the gold standard of scientific evidence because judges, jurors, the public, and conscientious defense lawyers are unaware of, or discount, the false positive probabilities that are its dark side.[794] The idea that DNA evidence is inherently backed by rock solid science has been aided by prosecutors, and their expert witnesses who have been allowed to testify with minimum critical restraint by judges.[795]

In spite of popular mythology about its accuracy, there is likely a high false positive rate in matching DNA samples.[796] That is hinted at by the results of a controlled test conducted under laboratory conditions, involving

72   Menace To The Innocent

DNA samples obtained from 225 FBI agents, 14 months apart.[797] The test excluded the possibility of a variation in the quality of the samples or any other factor that could cause a statistical variation in matching a DNA sample with the correct person.[798] Yet using standard testing techniques and protocols, there was a 12-1/2% false positive rate – *one out of eight* DNA samples was matched to the wrong person – when the two sets of samples were matched by FBI crime lab technicians.[799]

DNA false positives are thus known to occur in both proficiency tests and actual cases.[800] Furthermore, false positives are also known to be caused by factors other than technician malevolence.[801] Among those are contamination caused by the method of collecting a sample and/or its mishandling and/or its storage before or after arriving at a lab, contamination caused by unsterile conditions in the laboratory environment, error in conducting a test, misinterpretation of a test result, and erroneous reporting of a test result.[802] The National Research Council (NRC) summarized those as laboratory errors "in sample handling, procedure, or interpretation."[803] Substantiating that the FBI lab's 12-1/2% false positive rate of mis-identifying its agent's DNA samples was not an aberration, the NRC observed in its report, *DNA Technology in Forensic Science*: "Laboratory errors happen, even in the best laboratories and even when the analyst is certain that every precaution against error was taken."[804] So it is known a slip-up anywhere along the line by a well meaning and qualified lab technician can result in a DNA false positive that incorrectly associates an innocent person with crime related biological evidence.

The effect of unavoidable errors is that it was conservatively estimated in *DNA Matches and Statistics*: "... a reasonable estimate of the false positive error rate is 1-4 percent."[805] Yet in spite of the hard evidence to the contrary, the mythology of exactitude enveloping DNA evidence has reached the point that some "prosecution experts have suggested that false positive error is impossible in DNA analysis."[806] The reality of cases such as Josiah Sutton's proves the baselessness of that opinion.

It is almost redundant to point out that incorrect lab results have no relevance to a person's innocence or guilt, and the unreliability of test results under the best of conditions indicates the shaky evidentiary value of DNA related expert testimony.[807] That emphasizes the hazard posed to a suspect or defendant by conditions at the FBI and other crime labs that are ripe to escalate the likelihood of a false positive.[808] The probability of that increases to a certainty when anywhere along the line a lab technician maliciously contaminates or substitutes a DNA sample, alters a test or written report, or simply gives perjurious testimony about a test result to aid the prosecution in proving its theory of the case.[809]

Furthermore, whether a crime related DNA sample is deliberately or accidentally cross-contaminated with that of a suspect, retesting is of minimal practical value at discovering the existence of a false positive: Since the more conscientiously a retest is performed, the more likely it is to confirm the initial false positive attributable to contamination.[810]

Another contributor to a false positive was explained by Professor Sir Alec Jeffreys the discoverer of DNA fingerprinting, on the 20th anniversay of his discovery.[811] He noted that the database of DNA profiles used in a criminal investigation only documents ten identifying markers.[812] Those markers are used as points of comparison between the DNA profile of the people in the database and a crime related DNA sample. Jeffreys noted that as the number of people included in the DNA databases has increased, ten identifying markers is no longer sufficient to avoid the reasonable expectation of a false positive.[813] Jeffreys expressed his concerns specifically in regards to the United Kingdom where there are DNA profiles of 2.5 million people in police databases, but they are also applicable to the United States, where the FBI also uses ten markers in its DNA database that as of June 2004, contained over 1.8 million profiles.[814]

The possibility of a false positive is also heightened by the still evolving understanding that contrary to popular belief, a person's DNA is not set in stone, but has a fluidity alterable by a number of conditions.[815] It can be affected by physiological and psychological influences that can not only cause aspects of a person reflected in their DNA profile to change over an extended period of time, but literally before one's eyes.[816]

Menace To The Innocent

Chapter 9

A Random Match Probability And False Positive Probability Are Divergent

The divergent nature of the random match probability and the false positive probability in a case is made crystal clear by analyzing how they relate to Josiah Sutton's case.

If the prosecution's lab technician witness had testified about the probability of a random match, he could have expressed his level of certainty by testifying the odds were *a billion to one* that Sutton's DNA didn't match that associated with one of the woman's attackers.[817] He further could have made things look bad for Sutton by testifying that prior to the test the odds could be considered *100 to 1* he was one of the attackers, because the circumstances of his arrest suggested he could have been involved. He could have further made Sutton's chances for acquittal appear very dark by testifying that there was only a *1 in 1,000* chance that the test result was erroneous.[818] When finished with that testimony the jury might have been willing to vote Sutton guilty without even leaving the jury box. After all, why bother with going through the motions of pretending to deliberate in the jury room with such overwhelming scientific evidence of his guilt?[819]

Yet appearances can be deceiving, and reality can be far different than the picture painted by testimony elicited by the prosecutor locked onto the goal of obtaining Sutton's conviction.[820] In response Sutton's lawyer could have had an expert, such as Professor Thompson, testify about the "posterior odds" of a false positive DNA match: Bayes' theorem is the accepted scientific method of calculating those odds.[821] If the estimates testified to by the prosecution's witness were accepted at face value, the defense expert would have been expected to testify the posterior probability was *1 in 10* that a match between Sutton's DNA and the crime scene DNA was the result of a false positive.[822] In other words, there was a *ten percent* chance the prosecution's random probability odds of a billion to one was the result of a test result erroneously implicating Sutton.[823] So instead of clinching the case against Sutton and damning him to a long prison sentence, the prosecution's DNA testimony could have helped acquit him.[824] Why? Because when analyzed using recognized scientific standards, the veracity of that testimony would have fallen far below the 99% certainty standard the Supreme Court has inferred is required to support a conviction.[825] If the prosecutor had known the defense was going to counter the lab technician's testimony with the scientific probability there was a 1 in 10 chance Sutton wasn't implicated by the DNA test, it is possible the technician's misleading DNA testimony wouldn't have been presented. It is also possible the prosecutors would have sought dismissal of the charges without a trial.

The lab technician in Sutton's case didn't testify to those probabilities, but he did testify that Sutton's DNA was "definitely" that of one of the two men who raped a woman after a car jacking.[826] It would seem that Josiah Sutton's lawyer would have been eager to reveal that the prosecution's seemingly damning DNA testimony was little more than a smoke and mirrors show designed to deceive the jury about how weak his case actually was.[827] Yet Sutton's lawyer didn't do so. Why? He quite simply may not have been aware of the direct relationship between the probative value of DNA evidence when evaluated in light of the applicable false positive rate; and even if he had known it, the judge may not have allowed testimony bringing it to the jury's attention.[828] In this regard it was pointed out in *How the Probability of a False Positive Affects the Value of DNA Evidence* (2003): "[N]o court has rejected DNA evidence for lack of valid, scientifically accepted data on the probability of a false positive."[829] In other words, while every jury must be provided with a random match probability to use as a bench mark in evaluating the value of DNA evidence, not a single court in this country requires that they be told of the far more important estimate of how likely the DNA testimony is based on a "false positive," i.e., an erroneous identification.[830] If the prosecution expert's estimate is based on a false positive it has zero evidentiary value – but enormous prejudicial value against the defendant.

The prosecution's reliance on false positive DNA testimony to convict Josiah Sutton is not unique. Six years earlier a Tulsa, Oklahoma jury convicted Timothy Durham of raping an 11-year old girl after being presented with testimony they were led to believe substantiated that his DNA matched that of the girl's attacker.[831] The power of DNA testimony to overwhelm the sensibilities of jurors is indicated by Durham's jury: They chose to believe the testimony of a lone crime lab technician over the testimony of *11 alibi witnesses* who placed Durham in *another state* at the time of the crime.[832] Durham was released in 1997 after serving four years of his 3,000 year sentence, when it was proven the DNA testimony that contributed to his conviction was based on a false positive test result – and that his nearly dozen alibi witnesses had been telling the truth.[833] Durham's exoneration raises the question that if he had been permitted to present evidence to the jury that his alleged DNA link to the crime was due to a false positive analysis, they might have acquitted him.

Chapter 10

Wrongful Convictions Are Cemented with False Positive DNA Testimony

The dark side of DNA evidence is its use in criminal cases is heavily weighted against aiding the innocent.[834] DNA testing can be an effective tool for uncovering misjustices. The promise and potential extent of its use for that purpose is revealed by the fact that through September 2004 over 150 men and women convicted of felonies have been exonerated by DNA testing in the U.S.[835] However it is reasonable to suggest in light of what is known about DNA false positives, that the cases of Josiah Sutton and Timothy Durham only hint at the likelihood many times more innocent people have been convicted as a result of DNA testimony than have been exonerated by DNA evidence: If, for example, only 10,000 people have been convicted nationwide on the basis of DNA evidence,[836] then based on a 14% wrongful conviction rate, the use of DNA evidence in courtrooms has generated 10 times as many wrongful convictions than it has overturned.[837]

Although the sacred place DNA testing has in the legal and popular consciousness isn't warranted, that doesn't prevent the prejudicial effect of the popular misconception of its infallibility from easily overwhelming evidence of a person's actual innocence – or the prosecution's lack of corroborating evidence. In Timothy Durham's case, for example, what turned out to be insubstantial expert testimony linking his DNA to that of a rapist was believed by the jurors over eleven alibi witnesses placing him in another state at the time of the assault.[838]

That situation is compounded by a deliberate false positive DNA test result such as occurred in Gilbert Alejandro's case.[839] Gerald Davis was also victimized by false positive DNA testimony.[840] Davis was convicted in West Virginia of sexual assault and kidnapping based on Fred Zain's testimony that a DNA test didn't exclude him as the source of semen found on the woman's underwear.[841] Two post-conviction tests of the semen excluded Davis as the attacker, and his conviction was vacated.[842] He was acquitted in December 1995 after a retrial.[843] Gerald Davis spent eight years in prison due to the false positive DNA testimony at his trial.[844]

William Harris was another innocent man convicted in West Virginia of sexual assault after false positive DNA testimony by Fred Zain tied him to semen left by the woman's attacker.[845] After failing to comply with three orders by a judge to turn over DNA evidence for post-conviction testing, the prosecution complied after a fourth order was issued in response to a contempt motion by Harris' attorney.[846] After tests from two different labs excluded William Harris as the source of the DNA, the prosecutor conceded he was innocent.[847] Harris' conviction for which he spent eight years imprisoned, was vacated on October 10, 1995.[848]

The seductive ease with which DNA evidence enables a conviction to be cemented by what is commonly accepted as irrefutable scientific testimony, invites prosecutors, police investigators, and crime lab personnel to engage in the same types of abuse as are common with other forms of scientific analysis.[849] Thus it would be consistent with human nature that the wrongful conviction rate in cases involving sacrosanct DNA testimony/evidence is higher than in cases involving other forms of evidence.[850] That supposition is at least in part supported by the results of the only known proficiency test the FBI administered to its DNA technicians.[851] It was reported by an FBI agent that all the technicians – except for possibly one – failed the *open* test.[852] To prevent the results from being "discoverable" by defense lawyers, the head of the lab's DNA Unit acknowledged to investigators for the Office of the Inspector General that he ordered the proficiency tests destroyed.[853] It is unknown how many innocent men and women were erroneously implicated in a crime by a false positive test result and subsequent prosecution favorable testimony by the FBI lab technicians who failed that test.

Chapter 11

Bite Marks, Hair Analysis, And Other Skeptical Forms Of Evidence

As the broad range of examiner proficiency tests previously cited indicates, the unreliability of prosecution expert testimony runs the full gamut of what could be expected to be presented in court against an accused person.[854] Whether prosecution evidence consists of such things as a paint sample, firearm identification, blood sample, ballistics or bullet sample, documents or handwriting, hair sample, latent earprint, fiber sample, bite mark, soil sample, latent fingerprint, footprint, DNA, voiceprint, or an SBS diagnose, expert testimony related to it typically has a believability factor comparable to an eight-year-old caught with his hand in a cookie jar denying he was fetching a cookie.

I – Bite Mark Analysis

In December 1991, 36-year-old Kim Ancona was murdered in a Phoenix, Arizona lounge.[855] Whoever killed Ancona bit her in the course of the crime.[856] In 1992 Ray Krone, an acquaintance of Ancona, was convicted of her murder and sentenced to death.[857] The jury relied on forensic dentist Dr. Raymond Rawson's testimony that the bite marks matched Krone's teeth. The Arizona Supreme Court reversed Krone's conviction because the prosecutor concealed a potentially defense favorable videotape concerning the bite mark evidence until just prior to the start of his trial.[858] Retried in 1996, Ray Krone was convicted for a second time based on Dr. Rawson's testimony.[859] However he was sentenced to life in prison instead of death.[860] Having steadfastly proclaimed his innocence because he was home asleep at the time of Ancona early morning murder, Ray Krone told *The Arizona Republic* in an interview after his second trial: "I was not there that night. [This] pretty much rules out any faith I have in truth and justice."[861]

After visiting him in prison after his second conviction, a cousin that was a successful businessman hired a lawyer for Krone who began digging for evidence to clear him.[862] After spending about $300,000, the cousin's faith in Krone's innocence paid off when on April 4, 2002 he was excluded as Kim Ancona's attacker by DNA tests of saliva and blood found on her clothes and body, and less than a week later he was released after ten years of wrongful imprisonment.[863]

The erroneous bite mark evidence used to convict Ray Krone is not surprising, considering the results of a study by the American Board of Forensic Odontologists (ABFO).[864] The *Los Angles Times* reported in April 2002 that 63.5% of bite mark test results were "false positives," and 22% were "false negatives."[865] So the bite mark analysts were right 14.5% of the time, or in about 1 out of 7 cases.[866] That is less than one-third the 50% success rate that could have been expected if they had skipped the pretense of conducting a test and simply flipped a coin. The results of the ABFO study were consistent with Arizona State University Law School Professor Michael J. Saks description of bite mark testimony as "classic junk science."[867] Since bite mark evidence has such a high level of inexactitude, it is questionable if it could survive a well-prepared *Daubert* challenge.[868] The same doubt could be expressed concerning hair analysis, handwriting identification, and other suspect forms of evaluating physical evidence.[869]

II – Hair Analysis

Hair analysis is a subjective art that is as much a junk science as bite mark anlaysis.[870] There are no legal standards of hair analysis, no national association guidelines for examinations, and no generally accepted professional requirements for what constitutes a match between two samples.[871] The reason for that state of affairs is simple: A person's hair may, or it may not exhibit identifying characteristics under microscopic examination, and that uncertainty extends to hair from different parts of the person's scalp.[872] Thus there is no assurance of duplicability of results by two examiners following the exact same process of the exact same evidence – which is at the heart of the scientific method.[873]

The inexactitude of hair analysis was demonstrated by a Florida appeals court's overturning of Rodney Horstman's conviction in 1988 for the 1985 rape and murder of Sandra Peterson.[874] Horstman's conviction was based on the expert testimony of Michael Malone, a hair and fiber examiner with what is now known as the FBI's Trace Evidence Unit.[875] Malone testified the probability a pubic hair found on Peterson's clothing wasn't Horstman's was "almost non-existent."[876] The appeals court ruled: "Although hair comparison analysis may be persuasive, it is not 100% reliable. ... Thus, we cannot uphold a conviction dependent on such evidence."[877]

Another case involving false positive hair analysis was Ricky Hammond's 1990 conviction of sexual assault.[878] Although his conviction didn't solely hinge on expert testimony his hair was consistent with that of the woman's assailant, it was the sole scientific corroboration of the prosecution's narrative of the case.[879] Post-conviction DNA testing of vaginal swabs from the victim excluded Hammond as her attacker, and his conviction was reversed.[880] Acquitted after a retrial, Ricky Hammond spent over two years wrongly imprisoned.[881]

Edward Honaker was another man victimized by a false positive hair analysis.[882] In 1984 he was convicted of seven counts of sexual assault, sodomy and rape after a trial in which the sole scientific testimony was: "hair found on the woman's shorts "was unlikely to match anyone" other than Honaker."[883] Honaker was sentenced to three life sentences plus 34 years.[884] Two items related to the attack were known to contain sperm: a vaginal swab from the victim and her underwear.[885] In 1994, two separate post-conviction DNA tests unavailable at the time of Honaker's trial excluded him as the source of the sperm on either item.[886] On October 21, 1994 Edward Honaker was pardoned by Virginia's governor after ten years of wrongful imprisonment.[887]

Charles Fain spent 18 years on Idaho's death row after being convicted on February 24, 1982, of raping and murdering a 9-year old girl based on an FBI lab technician's testimony that three microscopic hairs found on the victim were likely to have come from him.[888] The girl was killed in Nampa, Idaho, but at the time of the attack Fain was living almost 400 miles away in Redmond, Oregon.[889] Furthermore, in pursuing charges against him, the prosecutor ignored the state polygraph examiner's assessment that Fain was being truthful in claiming he was in Redmond at the time of the attack.[890] On August 24, 2001 Charles Fain was released from Idaho's death row directly to the street, when the charges against him were dismissed after a DNA test excluded him as the source of the very same hairs used to convict him 18 years before.[891]

David Vasquez' February 1985 second-degree murder conviction in Virginia involved not only false positive hair analysis as the sole scientific evidence, but a false confession to the crime by Vasquez.[892] Although he proclaimed his innocence, he was sentenced to 35 years in prison after accepting an Alford plea to avoid a harsher sentence if he went to trial and was convicted of first-degree murder.[893] In a purely coincidental stroke of luck for Vasquez, in 1988 another man was implicated in a series of rape/murders involving a *modus operandi* similar to the one he was convicted of committed.[894] Based on that information and an FBI report that credited Vasquez's convicted crime to the other man, the prosecution joined in a defense motion for the governor to grant an unconditional pardon.[895] On January 4, 1989 Virginia Governor Gerald Baliles pardoned David Vasquez and he was released after five years of wrongful imprisonment.[896]

John William Jackson's murder conviction procured by a combination of hair analysis *and* odontology was reversed by a Florida appeals court in August 1987.[897] A prosecution expert witness testified that bite marks on the wrist of the victim were consistent with John Jackson's teeth.[898] Another prosecution expert, FBI crime lab technician Michael Malone, testified that two hair strands found on the victim's pajamas were "positively" Jackson's.[899] However John Jackson's fingerprints were *excluded* as matching those linked to the murder, and other hairs from the murderer found at the murder scene clearly did not come from him.[900] Thus the appeals court determined that in spite of the prosecution's expert testimony, it was a "reasonable hypothesis that someone else committed the crime."[901]

III – Handwriting Analysis

Handwriting analysis is another form of junk science. Furthermore, its unreliability as scientific evidence has not only been known internationally since the early 20th century, but it played a central role in the establishment in 1907 of the Court of Appeal Criminal Division in Great Britain.[902] The following are six cases involving erroneous expert handwriting analysis.

Adolf Beck

In 1877 a man known as Lord Willoughby was convicted in England of defrauding several woman of leisure (prostitutes) and served six years in prison.[903]

After his release women of leisure again began complaining to police of being defrauded with bad checks passed by a man calling himself Lord Willoughby.[904] In December 1895 one of those women recognized him on a street in London and he was arrested.[905] The man protested his innocence, but he was positively identified by several of the defrauded women, as well as two police officers who were familiar with Lord Willoughby from his previous conviction.[906] The man was subsequently convicted in 1896 based on a combination of eyewitness testimony and expert testimony that the handwriting on the swindler's checks matched that of the accused man.[907] He was sentenced to seven years in prison, which he served in full.[908]

In 1906, three years after his release, eyewitness testimony and expert handwriting testimony was relied on to again convict the same man of fraud charges similar to the previous two convictions.[909] As the man was to begin his prison sentence as a habitual offender, a man calling himself Thomas was arrested on similar bad check charges.[910] After seeing Thomas, the women who had testified against the just convicted man realized he wasn't the man who had defrauded them: He merely resembled the actual con artist – Thomas – whose variety of aliases included Lord Willoughby.[911] The ensuing investigation determined the man called Thomas had committed all the crimes related to the three convictions – in 1877, 1896 and 1906.[912] However Thomas had only been involved in the first prosecution in 1877.[913]

An innocent man was convicted of Thomas' crimes in 1896 and 1906, based on a combination of mistaken eyewitness testimony by women who had seen Thomas face to face under conditions of zero stress, and experts testifying the handwriting on the checks matched that of the wrongly accused and convicted man.[914] That man was Adolf Beck, a Norwegian mining engineer. Beck was granted a "free pardon" and awarded £5,000 in compensation for having been wrongly convicted twice, and being imprisoned for over seven years.[915]

Adolf Beck's exoneration brought attention to two facts that were ignored during his first prosecution and imprisonment that cast serious doubt on his guilt. First, the jury ignored clear proof he had been in Peru when "Lord Willoughby" was imprisoned in 1877, and therefore he couldn't be him.[916] Second, he was able to get the trial court to reexamine the case when prison records from "Lord Willoughby's" 1877 conviction showed he had been circumcised, whereas Beck had not been.[917] However the court did not grant Beck a new trial and he served the full seven years.[918]

Public outrage over what happened to Adolf Beck in spite of the clear evidence of his innocence, led to the creation in 1907 of the Court of Appeal in Great Britain to review criminal convictions.[919] Prior to establishment of the appeals court a Royal Pardon was a prisoners only means of relief from a wrongful conviction.

Herbert Andrews

A case in this country reminiscent of Adolf Beck's began in the summer of 1913, when a large number of bad checks were passed in and around Boston, Massachusetts.[920] In November 1913, the police became aware a department store clerk was believed to have written a bad check for $30.[921] The man, Herbert Andrews, was picked up and interrogated at Boston police headquarters.[922] Eyewitnesses verified his identity from a photograph, and his fingerprints were also taken to cross-check with those on the bad checks.[923] He was subsequently charged with forty counts of forgery and passing bad checks.[924] A handwriting expert Herbert hired in an effort to prove his innocence informed him his testimony would only help the prosecution, because his handwriting matched the writing on the checks.[925] *Seventeen* eyewitnesses testified at Herbert's trial that he had written at least one bad check in the person's presence.[926] He was found guilty and sentenced to fourteen months in prison.[927]

After Andrews was convicted and imprisoned, it was discovered the handwriting on the bad checks was the same as that on bad checks that had been passed in Salt Lake City, a city that Andrews had never visited.[928] Furthermore, bad checks with the same handwriting continued to be passed in the Boston area.[929] When a man named Earle Barnes was questioned, he admitted writing the bad checks that Herbert Andrews had been convicted of passing.[930] Herbert was released from prison, and Barnes, whose handwriting indistinguishable from that of Herbert, was convicted and sentenced to prison.[931]

Ronald Winslow

Another well known case of erroneous handwriting analysis was when 14-year-old cadet Ronald Winslow was expelled from the Royal Naval College in 1908 for allegedly stealing a five-schilling postal order.[932] His father, Arthur Winslow believed in his son's innocence and hired one of England's most distinguished barristers, Sir Robert Morton to fight for his son's exoneration.[933] In 1910 Ronald was cleared when it was proven that Ridgley Pierce, England's foremost handwriting analyst, had mistakenly identified Ronald's handwriting as the same as the writing on the forged postal order.[934] The case is the basis of the 1946 play by Terrence Rattigan, *The Winslow Boy*, and two movies by the same title, the 1948 original, and the 1999 remake.

Alfred Dreyfus

An erroneous handwriting analysis by Alphonse Bertillon, one of the 19th century's great criminologists, led to a well-known case of wrongful conviction: Alfred Dreyfus' 1894 treason conviction in France.[935] Bertillon incorrectly determined that Dreyfus wrote a document passing French military information to the Germans.[936] Dreyfus spent five years on Devil's Island in French Guyana before evidence of his innocence led to his release and subsequent exoneration in 1906.[937]

Howard Hughes' autobiography

One of the most sensational cases revealing the unreliability of handwriting analysis occurred in 1972 when writer Clifford Irving presented to the world what he claimed was Howard Hughes' autobiography.[938] It was actually a fake that he created in collaboration with British writer Richard Suskind.[939] Yet it was recognized as authentic when world renowned handwriting experts certified the writing on documents related to the hoax as that of Howard Hughes.[940] The scheme was exposed when Hughes broke a 15 year public silence by holding a telephone press conference in which he denied ever meeting, much less having heard of Clifford Irving prior to news reports about his purported autobiography.[941] After Hughes' press conference Irving confessed Hughes' autobiography was a fake.[942] Given the handwriting experts authentication of documents related to the book, it is possible Irving's fraud would not have been exposed without Hughes' public denial.[943]

Hitler's Diaries

An internationally publicized case of handwriting forgery was Hitler's diaries. In April 1983 the West German magazine *Stern* published excerpts from what they claimed were diaries of Adolf Hitler written between 1932 and 1945. Gerd Heidemann, a *Stern* journalist, claimed he paid 9.3 million DM (German marks) (about US$3.7 million) for the 60 small diary books. The diaries had allegedly been smuggled from a crash site in Dresden by a 'Dr Fischer.'

After examining one page from the diaries, handwriting experts in Europe and the USA agreed the handwriting was that of Hitler, and the authenticity of the contents was corroborated by history experts. Doubts persisted about their authenticity, and *Stern* provided some of complete diary books to the German Federal Archives (Bundesarchiv) and Swiss experts for forensic analysis. The forgery was exposed when it was established the diaries were written on modern paper with modern inks, and the writing was less than a year old.

The actual author of the Hitler diaries was Konrad Kajau, a Stuttgart forger of Nazi memorabilia. Heidemann was in on the scheme. Heidemann was convicted of stealing 1.7 million DM from *Stern*, Kajau was convicted of receiving 1.5 million DM, and they were sentenced to 42 months in prison.

Menace To The Innocent

Postscript

Given the prevalence of erroneous identification as a continuing cause of wrongful convictions, and that handwriting analysis has not advanced in reliability during the past century, there is no inherent safeguard preventing a reoccurrence of personal catastrophes such as befell Adolf Beck, Ronald Winslow, Herbert Andrews and Alfred Dreyfus – in three separate countries.[944] However, a positive development is several courts have denied admittance of "expert" handwriting testimony on the basos it doesn't meet *Daubert's* criteria as scientific evidence.[945]

IV – Earprint Identification

Earprint identification is another form of junk science, the use of which contributed to David Kunze's 1997 wrongful murder conviction.[946] In December 1994 a man was murdered during the burglary of his Vancouver, Washington home.[947] The victim's 13-year-old son survived a beating by the assailant, and described him to police as "a darkly complected male, possibly Puerto Rican, about six feet tall with medium build, dark or black hair to mid-ear, 25 to 30 years of age, and a deep voice."[948]

Kunze was the ex-husband of the murdered man's fiancé.

In spite of being physically dissimilar to the attacker – he was in his mid-forties, wore glasses, had reddish-blond hair, and was not Puerto Rican – Kunze was considered a suspect. Furthermore, the 13-year-old knew Kunze and didn't identify him.[949] Corroborating the mismatch of his identity with that of the murderer, searches of Kunze's home and vehicle didn't reveal anything suspicious.[950]

However, investigators found what they believed was a latent earprint on a door in the burglarized house, and a Washington State Police Crime Lab technician, who had never before attempted to match a latent earprint with a control sample, compared it with multiple impressions taken of Kunze's earprint in September 1995.[951] In spite of his absence of experience with earprints, the technician, Michael Grubb, claimed: "David Kunze is a likely source for the earprint."[952]

In June 1996 Kunze was charged with aggravated murder, assault, robbery, burglary and kidnapping, and in December 1996 the trial court held a *Frye* hearing on the admissibility of the earprint evidence.[953] Twelve experts from around the country with considerable experience in forensic techniques testified to varying degrees of positiveness that the veracity of latent earprint identification was not generally accepted by the scientific community, and that the FBI does not use the technique.[954] However the trial judge ignored those witnesses and instead relied on two prosecution witnesses in ruling earprint identification was an accepted scientific technique, and that he would allow testimony Kunze's earprint matched the latent earprint found at the crime scene.[955]

Kunze was convicted in 1997 after the jury heard prosecution expert Van der Lugt testify that he was "100 percent confident" the ear print left at the crime scene was Kunze's, and he was sentenced to life in prison without the possibility of parole.[956] In 1999 the Washington Court of Appeals reversed Kunze's conviction and ordered a retrial in ruling: "general acceptance may not be found "[i]f there is a significant dispute between qualified experts as to the validity of scientific evidence." At the very least, this record shows such a dispute. Accordingly, we hold that latent earprint identification is not generally accepted in the forensic science community."[957]

During Kunze's retrial a prosecutor violated the trial judge's pretrial order that no mention could be made to the jury of Kunze's previous trial or successful appeal. In March 2001 the judge granted a defense motion for a mistrial.[958] Lacking any evidence of Kunze's guilt, the prosecution declined to try Kunze a third time.[959] After being wrongly convicted on the basis of junk science evidence and spending almost five years wrongly imprisoned, David Kunze told reporters: "That's a horrible thing to stand there and have someone say you are guilty and you're going to prison the rest of your life. We all think it can't happen to us."[960]

Insubstantial earprint testimony has also been used to convict the innocent in England. In December 1998 Mark Dallagher was convicted of smothering a 94-year-old woman based on the testimony of two prosecution ear experts, one who said he was "absolutely convinced" the defendant's left ear was identical with the earprint of the woman's murderer that was imprinted on the window through which he entered her home.[961] After Dallegher had been imprisoned for seven years, his conviction was quashed on January 22, 2004 based on his exclusion as the intruder by testing of trace amounts of DNA recovered from the earprint – the same earprint that had been used to convict him.[962]

V – Shaken Baby Syndrome

Another purported crime detection method – diagnosing Shaken Baby Syndrome (SBS) – is under intense scrutiny as possibly belonging to the realm of junk science.[963]

Hundreds of cases of SBS are diagnosed each year in the U.S., the U.K.

Menace To The Innocent

and other countries.[964] However it was reported in the *British Medical Journal* (*BMJ*) in March 2004, that a certain type of eye injury – "subdural and retinal haemorrhages" (severe bleeding into the eye) – that has been used to diagnose the syndrome, can have causes other than abuse.[965] Researchers at the Wake Forest University School of Medicine in North Carolina reviewed the medical literature and case studies on the syndrome, and found: "Statements in the medical literature that perimacular retinal folds are diagnostic of shaken baby syndrome are not supported by objective scientific evidence."[966] More precisely, they found there are only two flawed case-control studies on the subject, and that the published work displays "an absence of ...precise and reproducible case definition, and interpretations or conclusions that overstep the data."[967]

Although ideally bleeding into the eye is one of three criteria present in a child for a diagnose of SBS,[968] it has been used as the singular criteria to diagnose the existence of the syndrome.[969] However a number of doctors are questioning whether the syndrome actually exists, or is the artificial construct of doctors seeking a convenient explanation for a child's death.[970] Suggesting that the science underlying SBS is "uncertain," Dr. Jennian Geddes questions the concept of the syndrome: "We need to reconsider the diagnostic criteria, if not the existence, of shaken baby syndrome."[971]

The March 2004 report in the *BMJ* – *Perimacular Retinal Folds From Childhood Head Trauma* – is not the first questioning of the scientific basis of SBS: In a review of review of SBS medical literature from 1966 to 1998 that was published in 2003 by the *American Journal of Forensic Medicine and Pathology*, researcher Mark Donohoe found: "there was inadequate scientific evidence to come to a firm conclusion on most aspects of causation, diagnosis, treatment, or any other matters," and that in regards to SBS, there are "serious data gaps, flaws of logic, inconsistency of case definitions."[972] Indicative that SBS's origin had no scientific basis, and that its suspect beginnings have never been substantiated, Donohoe reported: "the evidence for shaken baby syndrome appears analogous to an inverted pyramid, with a very small database (most of it poor quality original research, retrospective in nature, and without appropriate control groups) spreading to a broad body of somewhat divergent opinions."[973]

Furthermore, the *BMJ* indicated doubts about the scientific basis of SBS in a March 27, 2004 editorial: "We need to reconsider the diagnostic criteria, if not the existence of shaken baby syndrome" – extend to the diagnosis of child abuse in general, "...lack of case definitions or proper controls can be leveled at the whole literature on child abuse."[974]

Yet in spite of its absence of an identifiable scientific foundation, an SBS diagnosis testified to by a prosecution expert(s), has been relied on by judges and juries to convict hundreds, if not thousands of people, any number of which – such as Sally Clark and Angela Cannings England, and Ken Marsh in the United States – are the innocent victim of a medical misdiagnoses.[975]

Chapter 12

Ill-Founded Expert Testimony Is A Godsend To Prosecutors

Prosecutors focus so much attention to ensuring a case is supported by expert testimony because judges and jurors consider experts to be the most persuasive witnesses.[976] The District of Columbia's Court of Appeals recognized that a "mystic infallibility" is ascribed to courtroom scientific testimony.[977] That attitude has become even more ingrained as an aura of certainty has developed around DNA typing since its invention by British geneticist Alec Jeffreys in 1984.[978]

Experts essentially testify as eyewitnesses for the physical evidence of the crime that can't speak for itself.

An expert's opinion about evidence can not only have considerable influence over the outcome of a trial, but it can also influence the plea bargaining process.[979] When the prosecution has expert evidence that is alleged to implicate a defendant, that person is much more likely to agree to a plea bargain in an effort to try and minimize his or her sentence.[980] That effect is borne out by the numerous cases where a defendant later proven to be innocent, pled guilty to avoid a harsher sentence if convicted.[981]

One of the more egregious cases where this is known to have occurred involved three people pressured to plead guilty to the alleged June 1999 murder of a newborn child in Chowtaw County, Alabama.[982] However contrary to the contention of the prosecution's witnesses, the alleged mother was medically proven after their convictions to have been incapable of becoming pregnant due to a successful tubal ligation, and that she did not give birth to the child.[983] Since the alleged child had never existed there was simply no medical basis for the three people to have been charged in the first place, much less threatened with the death penalty if they went to trial and lost: it was impossible for a murder to have taken place.[984]

Expert testimony can have a scientific basis on par with the horoscope on a newspaper's comics page, and the conviction of three people for the murder of a non-existent child – not in the Dark Ages, but in 2001 – is an extreme but all too real example of that reality.[985] So one of the law enforcement system's dirty little secrets is the prevalence of suspect crime lab tests and expert testimony that prosecutors pawn off on judges, jurors, defense lawyers, and the public as scientific.[986]

A representative example of that is a 1989 report concerning fingerprint evidence by a statistician and a forensic scientist that was not publicly released by the British Home Office until 1995 – six years later.[987] Why? The report concluded that the fingerprint point system that had been used to cement the conviction of untold thousands of defendants: "had neither logical nor statistical justification."[988] The report had far reaching implications because it undermined a method of identification used to convict people around the world for nine decades.[989] A similar tale of deception could be told of any number of procedures the public has been misled to believe have a scientific basis.[990]

The reason for this is no secret to law enforcement insiders and astute outside observers. As a former forensic lab technician frankly observed: "People say we're tainted for the prosecution. Hell, that's what we do! We get our evidence and present it for the prosecution."[991] That basic conflict of interest is compounded by the staffing of many crime labs, including the FBI's, with career law enforcement agents influenced by the agency's agenda in a case, not laboratory technicians dedicated to searching for the truth and letting the chips fall where they may.[992] Retired FBI lab metallurgist William Tobin describes the lab as "a "fraternity" – a "shoot from the hip" culture favoring the prosecution, rather than a culture of objective science dedicated to an honest search for the truth."[993]

In the wake of the OIG's report on the FBI lab, NACDL President Lefcourt observed: "These so-called experts are usually not even scientists. They are FBI agents *posing* as scientists in court, performing tests they are not qualified to perform, and offering 'expert' opinions, under oath and under penalty of perjury, that they are not qualified to give. For too long, these supposed experts have taken the witness stand, taken an oath to tell the truth, and then said whatever came into their heads so long as it favored the prosecution, unchallenged by the court and defense counsel."[994]

Furthermore, a lab technician told OIG investigators they "*are taught to testify...to favor the prosecution,*" when a more sophisticated test undermines the prosecution favorable result of a less reliable testing method.[995] The tainting of testimony to favor the prosecution was explained to the OIG's investigators by one of the few trained scientists in the FBI lab: "*... the court is not being presented with technical testimony that is reliable.* Biased both by the person's ego and/or biased by a feeling – and particularly

Menace To The Innocent

with the fraternity – that they feel that they develop with the Police Departments or whatever. One of the difficulties that I've seen through the years is too strong an association, and affiliation, and a fraternization with the boys in blue. *They are not detached scientists.*"[996]

In an article for a professional publication, law and forensic science Professor James Starrs clarified what the FBI lab's pro-prosecution culture means in actual practice: "They analyze material submitted, on all but rare occasions, solely by the prosecution. They testify almost exclusively on behalf of the prosecution. ... As a result, their impartiality is replaced by a viewpoint colored brightly with prosecutorial bias."[997] In an interview with an author of *Tainting Evidence*, Professor Starrs explained the consequences of the close association between crime lab technicians and the prosecution: "That is what I have come to call putting the cart before the horsing around. They're effectively running the investigation backward, starting with a hypothesis of guilt, then going out to try and prove it. *That is not science. These people aren't scientists.*"[998]

UC Irvine Law Professor William Thompson has expressed a complimentary point of view: "The culture of such places, run by police or agents, for police or agents, is often just inimical to good scientific practice. The reward system, promotion, incentives ... in the end your pay check is based on successful prosecutions, not good science."[999]

Similar concerns were emphasized by Centurion Ministries founder James McCloskey in *Convicting the Innocent* (1989): "We see instance after instance where the prosecutor's crime laboratory experts cross the line from science to advocacy. They exaggerate the results of hairs, fibers, blood, or semen in such a manner that is absolutely devastating to the defendant."[1000]

A crime lab's endorsement of suspect or possibly even false physical evidence as legitimate, or its concealment of exonerating evidence, is an integral part of its role as part of what has been described as the law enforcement assembly line that is designed to produce convictions.[1001] Stretching a few rules here and there to manufacture the appearance a person has criminal culpability is simply a part of the process necessary to keep the process running smoothly.[1002] The lack of actual evidence against a person is nothing more than an inconvenient nuisance to an experienced crime lab technician.[1003] It doesn't interfere with the person being falsely painted by the technician as the bad guy or gal, to fit the prosecution's narrative of the crime.[1004]

To that observation can be added what is considered the crown jewel of evidence and the modern atomic weapon of prosecutors: DNA evidence. There is nothing special about DNA testing that makes it immune from the same types of manipulation, contamination, technician error, incompetence, and fraud that is known to happen with every other form of physical evidence testing and evaluation.[1005] However what makes false or erroneous DNA testimony linking a defendant to a crime particularly prejudicial, is the popular lore that a DNA sample is inseparable from the person it originated from.[1006] Thus prosecution testimony related to DNA evidence can, and does, easily overwhelm juror doubt about a defendant's guilt.[1007]

Consequently, prosecution expert testimony needs to be critically viewed from the perspective it is not derived from a careful unbiased evaluation of the evidence, but that it is the product of a presumption of guilt considered to envelope the suspect.[1008]

The judicial processes function of reliably determining a defendant's guilt or innocence beyond a reasonable doubt is debased by the automatic credence given by judges, jurors, and defense lawyers to a report simply because it emanated from a laboratory with an impressive sounding name, and which is supported by the testimony of a smooth talking "expert."[1009] Prosecutors rely on expert testimony related to allegedly scientific evidence precisely because of the reverence shown to it, irrespective of its distorting effect of giving credibility to evidence of questionable value.[1010]

Prosecutors are not just keen on expert testimony because of its powerful effect on a jury's likeliness to convict a defendant, but because defense attorneys knowledgeable of that effect may advise defendants – regardless of their guilt or innocence – to make the best deal possible rather than go to trial and face a more severe sentence after a conviction.[1011] The impact of fingerprint evidence in driving this phenomena was observed as early as 1924.[1012] In that year it was written in *Fingerprint and Identification Magazine*: "we are impressed by the large proportion of cases in which criminals confess when they learn that the finger-print system is being used. ... In one large Midwestern city, criminal lawyers refuse to take cases in which finger-print evidence figures. They cannot afford to risk their reputations on cases which will surely find their clients guilty."[1013]

Thus casting an eye askance at all crime lab results and expert prosecution testimony is vital to meaningfully protect an accused person's presumption of innocence.[1014]

That critical evaluation is not just essential in this country, because the U.S. is not unique in the way prosecution biased expert witnesses undermine the judicial process.[1015] It is also an endemic problem in other countries.[1016] In *Convicting the Innocent* (2003), Canadian prosecutor Bruce MacFarlane described its international prevalence in the following terms:

"I have already discussed similar situations in England, Australia, New Zealand and Canada where, putting the matter most charitably, forensic scientists working in government or police-operated laboratories felt aligned with the prosecution, resulting in a perception that their mandate was to support the theory of the police."[1017]

An argument can be made that the veracity of testimony by any crime lab technician – in any country – should automatically be questioned, and the work they do in every case placed under the highest level of independent scrutiny even if no previous problems have been detected.[1018] The protection of the innocent and any pretense law enforcement is concerned with justice all but requires that no evidence by a forensic laboratory can, or should, ever be taken at face value.[1019] As investigative reporter Martin Yant observed in *Presumed Guilty*: "The case books list enough false convictions caused by faulty crime-lab analyses to raise serious questions about the tremendous trust now placed in them by the courts."[1020] If for no reason other than the increased use of DNA evidence, Yant's caution is more apropos today than when published in 1991.[1021] Yet it is because such warnings are invariably ignored that shady and ill-founded expert testimony is a godsend to prosecutors.

Chapter 13

Minimal Crime Lab Performance Standards Breed Slothful Conduct

The standards of performance and expertise of laboratories testing evidence in criminal cases should be as high as that expected of any scientific laboratory in the world. Yet crime labs are operated at a lower level of oversight and with less competent people, than clinical laboratories.[1022] The 'cowboy mentality' prevalent in forensic laboratories is contributed to by the lack of an established "peer review system" monitoring their procedures.[1023]

That attitude is further fed by the absence of *external* blind proficiency testing of crime lab workers, even though it is the most accurate method of ascertaining a technician's skill level under "real-world" conditions.[1024]

At its most basic, during a blind proficiency test the technician is unaware their processing of a sample is being used to assess their skill level.[1025] However the National Research Council (NRC) emphasized in its report, *DNA Technology in Forensic Science*, that mere blind proficiency testing is not enough to adduce a technician's competence level: To do that a test needs to mimic real-life conditions in every particular.[1026] Such tests, for example, must be randomly administered so a technician never knows if a case they are working on is real or an examination of their skill level, and it must involve samples that are "truly representative of case materials."[1027] The NRC report also noted the importance of using degraded evidence samples typical of a crime scene to adduce a technician's competency: It used the example that blood "… tests based on pure blood samples would probably underestimate an error rate."[1028] Furthermore, proficiency tests must not only be blind and based on real-world conditions, but to ensure their impartiality and representativeness of actual results, Professor R. C. Lewontin observed: "there must be frequent *independent* and *unannounced inspections* and tests."[1029]

Such testing is not only crucial to ascertaining a technician's skill level, but it was noted in an article in the *Journal of Forensic Sciences*: "External blind proficiency testing is said to be the best source of information about laboratory error rates."[1030] That was an affirmation of the conclusion in an

NRC report: "… laboratory error rates must be continually estimated in blind proficiency test."[1031]

Blind proficiency testing of crime lab personnel nationwide is estimated to cost a maximum of $3 million a year.[1032] In relation to the almost $150 billion spent on law enforcement in this country every year, that is significantly less than a veritable drop in the bucket.[1033] Yet the FBI has successfully thwarted efforts to require external blind proficiency testing of its technicians.[1034]

That is not surprising considering the observations of FBI insiders, such as Bill Tobin, who in 18 years with the agency rose to be the lab's chief metallurgist: "FBI agents are like gods to some people, and jurors figure they must know what they're talking about, yet most of them are not scientists. They are basically people the bureau gets off the street, trains them for a year and then calls them bomb experts." [1035] In the same vein it was noted in *The FBI's Junk Science*: "In case after case, lab examiners [give] inaccurate testimony, with little or no scientific basis, about trace evidence that could link a suspect with a crime scene. … In the complex world of mass spectrometers and atomic-absorption spectroscopy, who would know, for example, that it can't be said, at least not scientifically, that two different paint samples, two bullet fragments or two shards of glass matched or came from the same source?"[1036] Mere association with the FBI and the use of authoritative sounding language is sufficient to have judges and jurors accept such testimony at face value, however scientifically baseless it may be.[1037] Yet its prevalence raises the serious question that if technicians with the most prestigious forensic laboratory in the country are able to openly practice with such a low level of competence, what is happening in lesser crime labs? The scandals related to Fred Zain, Joyce Gilchrist, and Arnold Melnikoff are likely only the barest tip of the iceberg of what lurks in the background waiting to be exposed.[1038]

Blind proficiency testing emulating real-world conditions faces stiff institutional opposition: by effectively requiring that forensic labs employ genuine "scientists" with integrity to the truth, it would shake up the status quo by jeopardizing presently unchallenged insubstantial testimony.[1039]

Suspect expert testimony is also protected to some degree by the situation that forensic laboratories are not required to comply with a federal law enacted in 1967 and strengthened in 1988, that imposes on clinical labs minimum quality assurance procedures, licensing requirements, and standards for supervisory and technical staff personnel.[1040] Consequently, clinical lab technicians must have a quantifiable level of competence that forensic lab workers are not required to exhibit.[1041] Molecular biologist Eric Lander described this anomaly in the following way: "Clinical laboratories must meet higher standards to be allowed to diagnose strep throat than forensic laboratories must meet to put a defendant on death row."[1042]

The generally low competence level of crime lab technicians underscores the menace posed to untold numbers of possibly innocent people by the millions of DNA profiles that have been, and are continuing to be entered into CODIS – the Combined DNA Index System – administered by the FBI.[1043] It was reported in April 2003 that no "audits of the DNA profiles in CODIS … were being conducted at any level."[1044] The potentially negative implications of that are heightened by disclosure of a 2001 audit by the Inspector General that half the labs contributing data to CODIS are not in compliance with the FBI's DNA standards, which the FBI itself doesn't conform with.[1045]

It is also noteworthy that universal accreditation of crime labs will not eliminate, or possibly even reduce the slothful results of crime labs, because accreditation involves subscribing to a uniformity of procedures, which does not in and of itself determine an accurate outcome of any particular test.[1046] Just as the continuing problems at the FBI's forensic laboratory[1047] indicates the lab's accreditation by the ASCLD has done nothing to improve the quality of test results and technician testimony.[1048]

The minimal competence expected of crime lab staffers is an indicator that misjustices are not just commonly being perpetrated in the name of science, but under the circumstances it would be illogical for anything other than that to be expected.

Chapter 14

The Subjectivity Of Forensic Evidence

Mental images of precise and exacting procedures are conjured up by the phrases 'scientific testing' or 'scientific testimony' based on a "scientific process."[1049] Yet insofar as forensic labs are involved, there are at least two dichotomies between those visions and reality. First, to be scientific a forensic analysis must satisfy a "scientific methodology ... based on generating hypotheses and testing them to see if they can be falsified; indeed, this methodology is what distinguishes science from other fields of human inquiry."[1050] Second, for testimony to be 'scientific' it must be based on a test result of physical evidence that is independently reproducible in other places and times.[1051] Thus, if physical evidence that results in a positive test result can be re-examined to result in a negative result a single time, or the positive result cannot be reproduced (such as inconclusive results), then it is a false positive with nil probative value.[1052]

The inability to reproduce a test result is one indicator that the initial result was not scientific in nature.[1053] Thus a contrary result is a much more important outcome than a confirmatory one, because it undermines support for the scientific veracity of the initial result.[1054] However, in spite of its significant value, crime laboratories do not as a matter of protocol submit evidence to an independent forensic laboratory to cross-check its results.[1055] Consequently, there is no way to know with positivity in a given case when prosecution testimony related to physical evidence is objectively based, or it is the unverified and possibly unfounded opinion of a lab technician.

The subjectivity dominating the forensic process is perhaps most clearly explained by the Clever Hans phenomena.[1056] In the early 1900s a horse in Germany named Hans "aroused world-wide interest," by his ability to correctly answer a broad range of questions.[1057] Among Hans' talents was accurately answering complex math addition, multiplication, subtraction, division and fraction problems.[1058] His expertise also extended to correctly giving the date for each day of the month, "he could tell time to the minute by a watch," he could recognize a person from a photograph taken many years before, and he understood different German dialects.[1059] He even exhibited the seemingly mystical power of reading a person's thoughts.[1060] In response to a question spoken to him by a questioner, Clever Hans answered by "tapping with his right forefoot."[1061] Hans caused such a stir because in some of his public demonstrations he had an accuracy rate of 100%.[1062] He also gave correct answers when people other than his owner, Wilhelm von Osten, questioned him.[1063]

In 1904 The Hans Commission was formed in Germany to determine how it was possible for a horse to correctly answer questions that required memory and complex thought processes for a human to do so.[1064] The commission was comprised of thirteen men with expertise covering various aspects of science.[1065]

After a number of tests the secret of Clever Hans' talent was discovered by the commission: He started and stopped tapping based on nearly imperceptible head movements of the person that asked a question.[1066] Hans the horse was very clever indeed, but not due to his knowledge or computation skills, but because he was a master at reading the 'tells' unconsciously signaled by his questioner.[1067] One consequence was Hans mirrored the questioner's answer to a question, whether it was erroneous or correct.[1068] Another indication that Hans read "tells" was the closer the questioner stood directly in front of him, the higher the rate of correct answers he provided.[1069] At the other end of the spectrum, if the questioner stood outside of Hans' field of vision, he did not answer.[1070] However Clever Hans was able to answer correctly if the questioner's back was turned, because it was head movements that provided the "tells" for how he answered.[1071] It is an indication of how much people wanted to believe Hans had extraordinary cognitive skills, that most observers, as well as von Osten, assumed the reason he answered correctly more often when a questioner was close to him was because he could *hear* better – when it was actually because he could see better to detect the head movements he interpreted as signals to start and stop tapping.[1072]

Hans' marvelous talent was *pantomiming* nearly imperceptible signals from the person seeking a response from him.[1073] The subjective process Clever Hans used was overlooked for so long by so many intelligent and perceptive people because he intuitively mimicked the same method master illusionists use to fool an audience – misdirection: While *the audience* was watching the tapping of Hans' right hoof, *he* was intently watching for 'telling' head movements by his questioner.[1074] Hans' hoof was a prop in the same way a woman sawed in half is a prop for an illusionist. Hans' success was indicated by the number of audience members who believed the tapping

Menace To The Innocent

of his hoof was in response to his brain's rapid calculating of dates, numbers, and other bits of relevant information. All that would have been needed to expose Clever Hans was a skilled 'illusionist,' would have been to put a blinder on him during a public demonstration because he could not accurately answer unless the questioner was clearly visible.[1075]

A form of the Clever Hans phenomena occurs daily in crime labs all across the country: The correct answer is 'telegraphed' to a technician by whoever is seeking the evidentiary value of suspected crime related evidence.[1076] The correct answer is typically presupposed by the question the technician is asked to answer, such as if a latent fingerprint matches that of a suspect.[1077] Yet as explained in a preceding section – *Fingerprint Analysis: Voodoo Palmed Off As Science* – a small area of virtually anyone's fingerprint can be similar to anyone else's fingerprint, so the presupposed answer of a match can successfully be provided.[1078]

A particularly blatant example of 'telling' the right answer to crime lab technicians was reported nationally in 1999. As part of the government's defense in *Mitchell* to a *Daubert* challenge to fingerprint analysis as a scientific process, the FBI requested that crime labs around the country examine the 10-print card of the suspect with latent crime scene prints for a match.[1079] Nine of the 34 crime labs responded with a negative match.[1080] The 26.5% rate of non-unanimity was unacceptable to the FBI, because the test was intended to support that fingerprint examination is a scientific discipline, with absolute certainty in its results.[1081] To elicit unanimity in the test results, the FBI 'telegraphed' its desired response to the nine non-conforming labs by requesting they reevaluate their determination with the aid of greatly enlarged fingerprint photos, on which the FBI marked suggested points of similarity with "red dots."[1082] The FBI also telegraphed its desired response by reminding those labs how important it was to report that all the crime labs unanimously matched the prints, because the results would be included in the prosecution's response to the defense's *Daubert* challenge to allowing the admittance of fingerprint evidence as scientific.[1083]

It is significant that the FBI only elicited a 73.5% positive match rate from what amounted to an open book test with the lab technicians only having a yes/no option of whether she or he believed the suspect's fingerprints matched the latent prints.[1084] If the test had been a double-blind test of say ten different, but similar fingerprint sets to compare to the latent prints, it is possible the number of responses identifying the suspect's prints may have only been 10% – or one out of the ten choices available to the examiners – which would have been approximately the same result as relying on random selection without bothering with the formality of an examination process.[1085]

The subjectivity inherent in the fingerprint examination process is not only reflected in the *Mitchell* test, but in the proficiency tests from 1983 to 2001 that had an adjusted average false positive rate in excess of 16%.[1086] That subjectivity is also expressed in texts about the identification process involving a latent print "whose clarity, or lack of it, makes a positive pronouncement of identity difficult."[1087] The matching of a suspect's print with a latent print may then depend on the examiner becoming convinced after looking at the latter print, that the former one has the same characteristics.[1088] This transference of characteristics has been described in the following way: "If the fingerprint technician should look at the known print before the unknown one, he is bound to be influenced, whether he realizes it or not, by the ridge formations in it. ... He may unconsciously see in the unknown print formations which were clearly visible in the inked print but which may not be clear in the latent impression. We do not mean to infer that he would do this with any dishonest intention. It is just human nature to do so, and may even occur without the technician's knowledge."[1089] A similar tainting effect occurs in every other forensic lab process – none of whose conclusions are objectified by the cross-verification of an outside testing procedure. The real world consequence of these subjective processes is reflected in the extraordinarily high error rate in crime lab technician proficiency tests, that in some cases exceed the odds of chance,[1090] and in the extreme divergence of opinion by fingerprint examiners in the FBI's test in the *Mitchell* case.[1091]

The tainting of a crime lab technician's judgment by transference of a suspect's known characteristics to crime related evidence is no different in principle than the false identification of an innocent person in a line-up or photo array by implanting the suspect's image in the witnesses mind.[1092] This phenomena is so well known that many jurisdictions have implemented safeguards in an effort to minimize such catastrophes.[1093] Yet no such protection of the innocent exists to protect them from being untowardly linked to a crime by their initial suspected association to crime related evidence. Furthermore, there is no such effort to even cursorily objectify the testing of physical evidence by forensic laboratories. This situation is worsened by the observation in *The FBI's Junk Science* about, "the institutional culture within the FBI that has long condoned inadequate forensic work and the bending of testimony to help the prosecution. The pattern is so well entrenched ... that agents don't have to be told to slant their testimony, they just know what is expected of them."[1094]

The ultimate example of the subjectivity of forensic examiners is when *they* disagree about the probative value of the evidence used to convict an innocent person. That was what happened when Danny McNamee was wrongly convicted based on erroneous fingerprint testimony by Scotland Yard examiners whose work had been *triple-checked*, but which was later exposed as unreliable by a consensus of fourteen fingerprint examiners.[1095] It also happened to John Stoppelli, as related in a Spring 2003 issue of *Justice Denied* magazine:[1096]

> John Stoppelli's tale of woe is told in *Never Plead Guilty*, the 1955 biography of the late San Francisco defense lawyer, Jake Ehrlich.
>
> In the late forties four men were arrested and twelve envelopes of heroin seized during a raid on an Oakland, California hotel room. A "ring" fingerprint was found on one of the envelopes. After searching through the Bureau of Internal Revenue's (now IRS) national fingerprint database of known criminals, an agency "expert" decided it matched the print of John Stoppelli, a New York hoodlum.
>
> That identification conflicted, however, with the failure of the four arrested men to implicate Mr. Stoppelli, and his claim that on the day of the raid he had registered with his probation officer in New York City, 3,000 miles from Oakland. He was nevertheless indicted for drug trafficking.
>
> Although the only evidence linking him to the drugs was the fingerprint "expert's" testimony implying the odds were six billion to one the "ring" fingerprint found on the envelope wasn't his, the jury found him guilty on the first ballot. At his sentencing, Jake Ehrlich argued for leniency, citing the evidence of John Stoppelli's innocence, and by stating: "I say he [the fingerprint expert] made a mistake. I say he is not the expert we were led to believe." The judge was unmoved and sentenced Mr. Stoppelli to six years in federal prison.
>
> John Stoppelli lost all of his appeals, even though he had obtained affidavits from the four men arrested when the envelopes were seized that he had nothing to do with the drugs, and the

Supreme Court refused to review his case.

As a last resort, Jake Ehrlich called in a favor and the FBI laboratory compared the lone fingerprint on the envelope with John Stoppelli's. It reported that although they had similarities, they didn't match. His claim of innocence vindicated, Mr. Stoppelli sought a new trial. His request was denied on the ground the FBI report wasn't "new" evidence, but was a reevaluation of the "old" fingerprint evidence. Stymied by the court's refusal to throw out his wrongful conviction, John Stoppelli sought, and was granted a commutation of his sentence by President Truman after he had served two years of his sentence.[1097]

John Stoppelli's case is just one of many that illustrate there is no bar to the subjectivity dominated forensic lab environment – that includes a technician's prosecution favorable testimony in a criminal case – from having a false positive rate comparable to that of Clever Hans when a blinder obscured him from picking up the telegraphed cues of his questioner.[1098]

Chapter 15

Prosecutor's Fallacy Skews Considering A Defendant's Possible Innocence

The prejudicial effect of unreliable forensic lab tests and testimony is compounded by a tactic prosecutor's have learned can effectively mislead jurors into believing the prosecution's evidence is much stronger against a defendant than it actually is.[1099] That tactic is to present damning expert testimony against a defendant that the expert claims is unlikely to have implicated the defendant if she or he wasn't the culprit.[1100] Known as the *prosecutor's fallacy*, the success of the argument is dependent on the absence of testimony explaining the likelihood that the damning evidence is nothing more than *straw evidence* that typically rests on the quicksand of a false positive analysis.[1101]

The prosecutor's fallacy occurs during closing arguments when the prosecutor emphasizes to the judge or jury that the positive identification of the defendant by the prosecution's expert or eyewitness(es), means there is an equally small likelihood the defendant is not guilty.[1102] Thus the expert's testimony is expected by the prosecutor to wipe out the possibility of reasonable doubt in the mind of jurors.[1103] Yet the commonness of false positives by prosecution experts and eyewitnesses all but obviates it as anything other than advisory – and not conclusive – evidence for a jury to consider.[1104]

An illustration of this is to assume the jurors in a rape case are told by the prosecution's expert witness that the odds are an overwhelming *billion to one* the defendant's DNA coincidentally matched that from the crime scene.[1105] However the jurors aren't told any probability related to the likelihood that the defendant was implicated by a false positive test result or evaluation.[1106] In his closing the prosecutor argues that since the likelihood of a random/coincidental match is infinitesimal, then the likelihood defendant is not the rapist is likewise infinitesimal.[1107] If the jury buys that argument, and in the absence of testimony related to the likelihood of a false positive they most certainly will, the defendant stands virtually no chance of acquittal.[1108] The jury will not be swayed that there is a reasonable doubt of the defendant's guilt, even if 11 credible people testify the defendant, who was personally known to them, was in another state at the time of the crime: Which is what happened in Timothy Durham's case.[1109]

Josiah Sutton's case is an example that as imposing as a billion to one odds of a coincidental match seems on the surface, it can easily translate into a 10% probability that an innocent defendant was falsely implicated by expert testimony related to a DNA test.[1110]

Sally Clark's November 1999 conviction of murdering two of her infant sons 14 months apart illustrates that the prosecutor's fallacy can also rely on statistics in non-DNA evidence cases to convince juror's of an innocent defendant's guilt.[1111] A successful lawyer, Sally Clark proclaimed her innocence, claiming both her children had suddenly stopped breathing.[1112] To undermine her claim, the prosecution presented expert testimony that the odds of a second baby in a family dying from "cot death" (Sudden Infant Death Syndrome – SIDS) after a previous death due to the syndrome is one in 73 million.[1113] Although the prosecution repeatedly emphasized to the jury the statistical improbability of two deaths in one family due to "cot death,"[1114] Clark's lawyer inexplicably, did not object to the statistical claim.[1115]

While pursuing her vindication, Sally Clark's husband discovered evidence that two cot deaths in one family is so common that one of every 50 families in the U.K. experiences a second cot death after losing a first child to the syndrome.[1116] So the actual probability of a second death was calculated by Mathematics Professor Ray Hill of Salford University to be approximately 1 in 95.[1117] Thus two cot deaths in the same family is *768,421 times more likely to occur* than her jury was told, and it reveals the prejudicial effect of the deceptive and incomplete expert testimony.[1118]

However, the persuasive evidence undermining Sally Clark's conviction was the discovery the prosecution had concealed hospital medical records confirming the death of her second son Harry was due to a viral infection.[1119] The concealed evidence also provided substantiation for the initial diagnosis that her first son, Christopher, died from a lower respiratory tract infection.[1120]

On January 29, 2003 the Court of Appeal quashed Sally Clark's conviction and she was released after more than three years of wrongful imprisonment.[1121] Sally Clark was not only victimized by the prosecutor's fallacy, but she was doubly victimized because the prejudicial argument about the one in 73 million probability of two "cot deaths" in one family that the jury considered in finding her guilty, had *nothing* to do with the death of her children.[1122] Unfortunately, other innocent parents have been victimized by the same junk science testimony underlying the prosecutor's fallacy in Sally Clark's case that two children dying in her family was so unusual that she must have murdered both.[1123]

The principle underlying the prosecutor's fallacy applies to many different types of evidence.[1124] The power of expert testimony in regards to physical evidence is emphasized by the weight given by judges and jurors to an eyewitness who testifies about the visual evidence of what she or he saw, and whose testimony as a *de facto* expert about what she or he did or did not see, is relied on to convict the accused person.[1125] Although eyewitness identification of a stranger is known to be very unreliable under a wide range of conditions and circumstances, it is persuasive when a witness points at the defendant and says: "He did it!"[1126] The defense's response may be identical to what it is with the testimony of a crime lab technician serving as the eyewitness for physical evidence: the defendant isn't the person the witness thinks they are identifying, because she or he wasn't at the scene of the crime and is innocent.

It is difficult to overstate the magnitude of how misleading the prosecutor's fallacy is about a defendant's possible guilt or innocence.[1127] However it is hinted at by the fact that in its *1992* report *DNA Technology in Forensic Science*, the National Research Council recommended that to minimize the prejudicial impact of the prosecution expert's random probability estimate on jurors and to help ensure they make an intelligent determination of the evidence, that they also need to be told of the false positive rate.[1128] Although better than the absence of any curative effort, there are concerned experts who don't think that is enough to offset the prejudice of the prosecution expert's testimony. UC Irvine Law Professor William Thompson has written: "Some experts have gone so far as to suggest that jurors be told *only* the false positive rate; they reason that the probability of a false positive is so much greater than the probability of a coincidental match ... that the later probability has little bearing on the value of the evidence."[1129] Professor Richard Lempert is one of those experts who recognize the prejudicial effect of testimony concerning coincidental match probability – it sounds impressive but has little probative value.[1130] Professor Paul J. Hagerman has also noted the rate of a coincidental match is irrelevant because it is dwarfed by the probability it is due to a false positive.[1131]

Due to the inevitableness of its prejudicially confusing effect on jurors, Professor Lempert has written: "jurors ordinarily should receive only the laboratory's false positive rate as an estimate of the likelihood that the evidence DNA did not come from the defendant."[1132] If the judge and jurors were informed of the expected false positive rate underlying the expert testimony of a crime lab technican or other specialized prosecution witness, or the likelihood an eyewitness was mistaken, the prosecutor's fallacy would be an ineffective argument and go the way of the Dodo bird – because its ability to sway the judge or jurors to convict a defendant depends on them being kept ignorant as to the true value of a prosecution witness' testimony.[1133]

Chapter 16

Are Prosecution Experts Criminals?

I s it criminal for prosecution expert witnesses to testify falsely as to the probative value of evidence?[1134] Although it is a question asked of an all but ignored problem, it is one of the single most important questions facing the law enforcement system today. Why? Whether a judge and jurors are mislead by testimony of a technician who stretches what s/he believes to be true, makes up testimony, withholds cautionary information, or testifies falsely about a test result, the result is the same: the truth finding function of the adjudication process is undermined.[1135] Yet the law enforcement system as a whole (and the courts in particular) only has legitimacy as a truth finding mechanism to the degree that it accurately does so.[1136] That is reflected in the stated purpose of the Federal Rules of Evidence: "That the truth may be ascertained and the proceedings justly determined."[1137]

Consequently it is a matter of concern to at least consider the potential criminal liability of prosecution experts who undermine the system's truth finding function by their illicit testimony concerning the value of evidence in this country's state and federal courtrooms.

Since the FBI provides expert evidence analysis for the prosecution in federal cases (although it also does so in many state cases) this abbreviated analysis will focus on issues related to the performance of their lab technicians. Insofar as their testimonial performance is concerned, their most obvious possible crime is perjury.[1138]

The federal crime of perjury is set forth in 18 U.S.C. §1621, which states in part: "Whoever – (1) having taken an oath before a competent tribunal. ... willfully and contrary to such oath states or subscribes any material matter which he does not believe to be true; ... is guilty of perjury."

The keyword concerning the falsity perpetrated by a person under oath in Section 1621 is it must be done *willfully*. The Supreme Court has generally recognized that willfully means a person acted with a "voluntary, intentional, and bad purpose to disobey the law."[1139]

Since perjury is not a strict liability statute,[1140] but requires proving the mental element of willfulness, there are at least five prongs related to the problem of initiating a perjury prosecution against a forensic technician.

One prong is the bureaucratic "code of silence" makes it as unlikely as a summer snowstorm in the Sahara that one crime lab member will testify against another,[1141] which can be essential to establishing criminal intent.[1142] An important component of looking the other way, is the discouragement of whistleblowing by fellow members of the FBI's crime lab brotherhood.[1143] This has been the general policy throughout the FBI's since J. Edgar Hoover let it be known he would not tolerate "... any whistleblowers in the FBI."[1144] On the hierarchy of whistleblowing, the most serious violation would be providing evidence a fellow crime lab worker's testimony was willfully untruthful.[1145] The "code of silence" protecting a technician from accountability for suspect testimony is augmented by the FBI's failure to internally hold technicians accountable by disciplining him or her for crime lab misconduct.[1146]

A second prong is many technicians, including those in the FBI may not keep detailed notes of their laboratory procedures.[1147] Those notes are either not made in the first place, or destroyed prior to or after conclusions are reported, precisely so they cannot be obtained by a defendant through the evidence discovery process.[1148] In the absence of documentation there is no way way to analyze and verify the actual methodological process used to arrive at the conclusion stated in a final report.[1149] This lack of documentation can shield a technician from criminal charges by obfuscating their intent to commit perjury with their false testimony, and possibly obstruction of justice by concealing the actual, or potentially exonerating nature of their test(s).[1150]

A third prong of the problem is the initiation of a prosecution when it can be proved, since many people think forensic technicians are protected from perjury charges by their critical connection to the prosecution: A former president of the National Association of Criminal Defense Lawyers (NACDL), Bill Moffitt, noted in this regard: "Only under the guise of the FBI could it not be considered perjury."[1151] In November 1997, after release of the OIG's report on the FBI lab, then NACDL president Gerald Lefcourt was more expansive in echoing the same assessment of the criminality of FBI lab technician testimony:

> "These so-called experts are usually not even scientists. They are FBI agents *posing* as scientists in court, performing tests they are not qualified to perform, and offering 'expert' opinions, under oath and under penalty of perjury, that they are not qualified to give. For too long, these supposed experts have taken the witness stand, taken

110   Menace To The Innocent

an oath to tell the truth, and then said whatever came into their heads so long as it favored the prosecution, unchallenged by the court and defense counsel. When an expert from the FBI lab was called to testify, his testimony might as well have come from the Burning Bush."[1152]

However regardless of a crime lab technician's perjury, it is in the self-interest of prosecutors who are aware of it, to "overlook" the prosecution witness' crime(s), because to not do so would undermine the basis of the conviction they are seeking to obtain with the aid of the perjured testimony.[1153] Thus prosecutors who are entrusted with ostensibly upholding the law, freely allow it to be criminally violated by a prosecution witness providing expert evidence considered crucial to securing a conviction.[1154]

A fourth prong of the problem with charging a forensic expert with perjury is determining if she or her *willfully* made the materially untrue statements implicating a defendant's association with alleged crime related evidence.[1155] Fredric Whitehurst, who at one-time was supervisor of the FBI's explosives lab, was blunt in his assessment that some of his former colleagues have the requisite intent to be prosecuted for perjury: "Some of these guys are liars."[1156] Yet on the other hand, some technician's are so lacking in competency that it is problematic whether they know when they are telling the truth or lying in court.[1157] A former chief metallurgist of the FBI's lab was quoted in *The FBI's Junk Science*: "In the case of [Robert] Webb and other agents ... there is a serious question of intent. ... Some examiners in the lab were so incompetent that were do you draw the line between knowing and unknowing?"[1158] It is difficult to think of a more damning indictment of the unreliability of the expert testimony routinely used to convict a defendant(s) – including untold numbers of whom are actually innocent – than that the forensic technician involved is too incompetent to have the intent necessary to have committed perjury or another other crime, such as obstruction of justice.[1159] The taint to testimony by a technician too ignorant of his subject to know it was untruthful, and hence to not have the requisite criminal intent to support a perjury charge, is compounded by some of the prosecution witnesses acting more like they are trying to pick-up a juror or someone else in the courtroom than give scientific testimony: This attitude was described by the same former FBI chief metallurgist: "I think it's more an effort to look good, to be the hot dog or the hero instead of being scientifically accurate."[1160]

A fifth prong of the problem with charging a forensic expert with perjury is that some of them are savvy enough to know the potentially criminal nature of their testimony: So the person will try to deliberately skirt around it.[1161] Forensic scientist John Thornton noted: "FBI agents have done this for years. They get around the issue of actual perjury by expressing an opinion: 'It's my opinion the paint came from the same batch.'"[1162] Yet in spite of the lack of an actual scientific basis or a general lack of expertise underlying those opinions, a technician is able to sway judges and jurors by the cachet of being associated with the FBI.[1163] That professional link deflects critical analysis of the cagey witness' untruthful and insubstantial – though technically not criminal – expert testimony.[1164]

To the degree that it occurs, courtroom criminality by forensic witnesses is a predictable activity: Since there is not a known case of an FBI lab technician's federal prosecution for perjurious testimony.[1165] This shield of protection from prosecution even extends to FBI lab personnel who *publicly admit* courtroom perjury.[1166] In 2002, FBI lab technician Kathleen Lundy admitted she had deliberately testified untruthfully during a pretrial hearing in a Kentucky murder case.[1167] Yet the Department of Justice declined to prosecute her related to her publicly *admitted* perjury that was nationally reported, and the FBI only suspended her pending an investigation.[1168]

Given that no FBI lab technician has been federally prosecuted for their perjurious testimony, it is almost superfluous to observe that no prosecutor, police investigator, or crime lab official has been prosecuted for the subornation of that perjury.[1169]

As long as a perjurious witness aids the prosecution's narrative of the crime, there is an attitude of tolerance by prosecutors for their criminal wrongdoing, and since they are the very people within the law enforcement hierarchy entrusted with enforcing laws such as perjury, it is apropos to ponder Juvenal's cautionary question – *Who shall keep watch over the guardians?*[1170] Insofar as crime lab personnel and other expert witnesses are concerned the answer is no one, since with rare exceptions they are able to violate the law with impunity.[1171] That condition, plus the absence of any significant political will to rectify that state of affairs, is indicative that there is no reason to believe there will be any change in the foreseeable future to that many decades long situation.[1172]

Yet that condition of non-criminal accountability is not neutral: As long as crime lab technicians are openly allowed to give untruthful or shady testimony by either perjuring themselves or cagily phrasing their answers to avoid perjury, the truth finding function of the courts of this country is subverted, and those in the know will have little or no faith in the veracity of that expert testimony, or a verdict dependent on it.[1173]

The foregoing discussion of perjury likewise applies to a prosecution expert retained by the prosecution who provides shady if not outright untruthful testimony.

Chapter 17

Double-Blind Testing Can Detect Inaccurate Crime Lab Tests

Analyst errors can so artfully be concealed by a crime lab's testing protocol that they would remain hidden from Diogenes if he could arise from the dead and wander the courtrooms of this country with his lantern seeking to establish the truthfulness of expert testimony.[1174]

A zero-blind technique is used by a crime lab that seeks to establish the similarity between a sample tied to a suspect (such as a fingerprint or DNA sample), and alleged crime related evidence.[1175] A zero-blind process is when an examiner and other people – such as other lab personnel and/or his or her supervisor(s) – know a suspect's sample is believed to be directly linked to, or have a common source as the crime related sample.[1176] So neither the technician nor those he or she has contact with is shielded from having a bias in the test's outcome.[1177]

The fundamental shortcoming of zero-blind testing is that it produces results significantly skewed towards confirming the expected result, such as finding that a tested item (such as a suspect's fingerprint) has a common source as the control sample (such as a latent fingerprint).[1178] The effect of a participant's knowledge on a test result has been established with the same level of certainty as the rotation of the earth around the sun.[1179] It has been confirmed e.g., by numerous medical trials that a placebo can produce results, including toxic side effects known to a participant that are similar to, or in some cases more pronounced than those caused by use of the actual drug.[1180] An example millions of people can personally relate to, is this anticipatory effect has been confirmed for the six most widely used anti-depressents in the United States: They failed to outperform placebo sugar pills in *over half* the 47 trials the U.S. Food and Drug Administration relied on to approve their sale to the public, and in the balance of the trials they only slightly outperformed the placebo.[1181] Affecting the outcome of a test by having one's anticipated result confirmed, is a form of self-fulfilling prophecy.[1182]

The prejudicial effect of zero-blind test procedures is so substantial that it can result in an error rate approaching 100%.[1183] What this means is that in a test of evidence allegedly originating with a suspect but which didn't, zero-

Menace To The Innocent

blind techniques can be expected to contribute to a crime lab technician declaring a match with the alleged crime related sample – irrespective of the fact they had different sources.[1184] So zero blind techniques allow for the mere suggestion that evidence is from a suspect to be sufficient for a technician's imagination to link it to the alleged crime evidence.[1185] This is not only true of initial tests – as was observed in the ASCLD fingerprint test results from 1983 to 1991, and CTS from 1995 to the present[1186] – but with "crime laboratories that do verify identifications [by] follow[ing] zero blind procedures exclusively."[1187] The authors of *Error Rates for Human Latent Fingerprint Examiners* explain the implications that has for the detection of possible technician mistakes that can also be attributable to zero-blind techniques: "The research is clear: under zero-blind conditions, if the first examiner has made an identification which is erroneous, the second examiner is likely to *ratify* the error rather than discover it."[1188] However, the uncorrectable deficiency of a zero-blind procedure to generate and then confirm an erroneous analysis has not interfered with its use as the technique of choice by crime laboratories.[1189]

Although single-blind techniques are known to produce more reliable results than zero-blind techniques,[1190] they are not used by crime labs for either initial tests, or follow-up verification tests.[1191] The main difference between zero and single blind techniques is that using the latter, both the technician and people she or he might have contact with, such as his or her supervisor and others associated with the prosecution, would know which sample was the suspects', while using the former, the technician would not directly be told that information.[1192] Withholding that information contributes to a single-blind procedure producing a more reliable result than a zero-blind technique, yet they both suffer from the same deficiency of not controlling conscious or sub-conscious influences on the technician to select the suspect's sample: Particularly considering that information can be telegraphed either deliberately or unknowingly by people who know which one it is – such as other lab personnel, a police investigator, or a prosecutor.[1193] That those messages can be telegraphed during zero and single blind tests by such "tells" as words or phrases, voice inflections, facial expressions, body movements, or gestures, is known with the same certainty as the moon's rotation around the earth.[1194] Furthermore that knowledge isn't "new," since the transmission of those tell-tale cues has been known for at least eight decades.[1195]

After the phenomenon that observers subtly and/or overtly send cues that influence the outcome of a zero or single-blind procedure became understood (circa 1930s), double-blind techniques began to be used to ensure the integrity of a scientific testing process.[1196] In contrast with the predictably tainted results generated by zero-blind and single-blind testing procedures, double-blind techniques produce scientifically reliable results.[1197] That is accomplished by eliminating any bias by people associated with the test toward the sample(s) that is (are) the object of the testing process.[1198] The *Random House Webster's Unabridged Dictionary* (1999) defines "double-blind" as: "of or pertaining to an experiment or clinical trial in which neither the subjects nor the researchers know which subjects are receiving the active medication, treatment, etc., and which are not: *a technique for eliminating subjective bias from the test results.*"[1199] In the context of a forensic laboratory, the elimination of "subjective bias" would be achieved by insulating the technician and *any* of the people she or he is in contact with from having *any* knowledge of which of multiple samples being compared to the crime scene evidence, is the suspect's sample.[1200] The "placebos" in a crime lab analysis would be decoy samples intermixed in the test array that would be as indistinguishable as possible from the suspect's sample.[1201] The technician would then have to examine the multitude of array samples to see if any matched the crime related sample.

The reason a "double-blind" test is so efficacious is that by isolating a result from the inevitable bias of the people involved, it can reliably separate a fictional (biased) insubstantial result from a factual (reliable) substantive one.[1202] Double-blind testing is a scientific process precisely because it undermines the "power of fiction" to influence its results.[1203] That is in stark contrast with the deficiencies inherent in zero and single-blind testing procedures that interfere with, if not prevent their achievement of results untainted by the "power of fiction": Which is why they are a form of quackery – not science.[1204]

The importance of using double-blind techniques to achieve reliable conclusions by differentiating fictional test results from substantive ones, is so well understood that they are "required for all federal drug testing programs, and for virtually all peer-reviewed, published scientific experiments."[1205] Yet not even prejudicial and inadequate single-blind techniques, much less infinitely more secure double-blind procedures, are required for forensic laboratory tests underlying a technician's possibly fictional courtroom testimony that can seal the fate of a defendant, whose innocence at that point is legally presumed.[1206]

The inadequacy of crime lab protocols to ensure the accuracy of test results underlying a technician's testimony is underscored by why the phrase 'blind test' was selected in 1937 to describe a scientifically controlled test.[1207] The phrase was inspired by a depression era advertising campaign that tested blind-folded smokers (multiple independent testers) to find if they could detect a particular brand of cigarettes from another brand (multiple samples).[1208] That ad campaign for Old Gold cigarettes was called "Take The Blindfold Test."[1209] It is another example of something almost too strange not to be true that an advertising company used a scientific methodology eight decades ago to test consumer preference for a cigarette brand that was light years more likely to produce an accurate test result than crime labs use today – in the 21st Century – to generate the experimental basis underlying a technician's expert testimony relied on by jurors (or a judge) to convict a person accused of a crime.[1210]

Furthermore, a typical consequence of deficient forensic test procedures is that a possibly innocent defendant spends years or decades of his or her life in prison: While in the most extreme cases it results in the execution of that person, whose presumption of innocence would have trumped the prosecution's case in the absence of the insubstantial expert testimony.[1211]

In spite of the fact that insulating a technician from untoward influences – such as subconscious 'Clever Hans' type "tells"[1212] – is known to be crucial to ensuring a scientifically sound unbiased test result, they are not followed by any crime laboratory in the country.[1213] That literally means that since every test conclusion and subsequent testimony by a crime lab technician relies on the "power of fiction,"[1214] it is automatically suspect as the insubstantial biased result of demonstrable defective zero-blind techniques.[1215] Under those egregiously unscientific operating circumstances, it does not take a Nostradamus or a Jeanie Dixon to prognosticate that errors by crime lab technicians have been the norm and not the exception up to the present time.[1216] It can likewise be predicted with certainty that the generation of erroneous results will continue to be the norm until *all* crime laboratories adopt double-blind techniques for *every test* they conduct. Although the need for all crime labs to institute the radical reform of exclusively using a scientific double-blind testing protocol can be described as self-evident, it is further emphasized by the fact that their current "cowboy" test procedures generate automatically suspect results even when a technician is conscientious and has no desire to deliberately taint the outcome.[1217] Double-blind test procedures also need to be mandatory for every non-crime laboratory that tests crime related evidence.

Chapter 18

Methodic Doubt Can Overcome
Pathological Science In The Courtroom

The types of physical evidence subject to testing by unreliable crime lab procedures are known, as are the ways prosecutors use expert testimony related to that evidence to mislead judges and jurors to obtain a guilty plea or a conviction after a trial.[1218]

Such prosecutorial tactics are able to flourish for one reason: Crime laboratories function as incubators of what Nobel Prize winning Chemist Irving Langmuir has termed "pathological science."[1219] He originated that phrase to describe "the science of things that aren't so."[1220] That description eerily sums up the overarching premise that pervades the operation of crime laboratories subtly, and not so subtly influenced to aid the prosecution's narrative of a case.[1221] The reliance of crime labs on deficient zero-blind techniques – a form of "pathological science" known to generate suspect and malleable results – is critical to that process.[1222] Although misstating the probative value of prosecution evidence is only successful to the degree that jurors and the public in general are deceived, its commonness was hinted at by Michael Braden, chief medical examiner of New York City for more than a quarter of a century: "Much as I hate to admit it, the sad fact is that some forensic scientists do, indeed, fool a lot of the people a lot of the time."[1223] Those people are fooled at times by a technician ignorant he or she is doing so, and at other times by a scheming technician.

Although the negative effect on a hapless defendant can be the same, the legal and moral implications of unintentional factors on the unreliable connection of a crime's physical evidence to the defendant is compounded to the ninth-degree by a forensic technician or free-lance prosecution expert who deliberately misstates the evidence's probative value in a report, in an affidavit, or in courtroom testimony.[1224] In such situations the evidence against a defendant is not due to its incriminating nature, but is solely attributable to what can be described as "wishful science,"[1225] or "voodoo science."[1226] A consequence of the knowing misstatement of an item's evidentiary value is that the jurors, the judge, and any other interested persons are fooled, and made fools of, by the prosecution's expert witness and anyone else involved in the scheme of deception – whether they be lab superiors, police investigators, and/or prosecutors.[1227] In every such instance a defendant is the victim of a frame-up, regardless of their guilt or innocence of the underlying crime they are accused of committing.[1228]

However, the prosecution's reliance on 'fake' evidence – i.e., evidence ascribed false probative value by the testimony of an expert witness – tends to cast a reasonable doubt on a defendant's guilt in the absence of compelling evidence to the contrary: Since if such evidence existed, it would have obviated the need for the technician's disingenuous testimony about that evidence.[1229] As physics professor Robert Park put it in *Voodoo Science*: "Most people who are drawn to voodoo science simply long for a world in which things are some other way than the way they are. Some cannot accept that we are prisoners of the Sun. They look wistfully at the stars that fill the night and imagine that there must be some way to overcome the limitations of space and time."[1230]

Malevolent motives however, are not required for unreliable crime lab test results and technician testimony to be presented as authentic in state and federal courts nationwide.[1231] Misplaced confidence in a testing procedure or one's competence can lead to courtroom testimony as unreliable as if it had been deceitfully given. Nobel Prize winning physicist Richard Feynman noted that in science: "The first principle is that you must not fool yourself and you're the easiest to fool."[1232] That principle is manifested in the reliance of crime labs on unscientific zero-blind testing protocols that at a minimum are susceptible to being slanted by the subconscious bias of the technician and other people he or she comes into contact with.[1233] The high rate of erroneous crime lab results covering a wide range of forensic examinations from fingerprints, to bullet analysis, to document examination, to DNA – that are attributable to such factors as flawed procedures, carelessness, or incompetence – is indicative that while they may exist, untoward intentions are not necessary for crime lab workers to fool themselves into believing that their conclusatory reports and/or testimony is reasonably based on what is actually the nil probative value of the prosecution's physical evidence to link a defendant to a crime.[1234]

The relevance of that evidence – irrespective of the intentions of the technician(s) involved – is further undermined by reliance of forensic labs on junk-science.[1235] That crime lab practice is encouraged by the admission of testimonial evidence by judges who fail to exercise their "gatekeeper" function of ensuring alleged "scientific" testimony actually has a scientific basis.[1236] The authors of *Tainting Evidence* described this situation thusly, there is "… growing concern about what has been termed "junk science" in U.S. Courtrooms. … The inability of courts to tell the difference between real and junk science was partially responsible for what seems like downright laxity when faced with the shortcomings of forensic examiners."[1237] The admission of junk science occurs on the federal level in spite of the Supreme Courts admonishment in *Daubert* that "under the Rules the trial judge must ensure that any and all scientific testimony or evidence admitted is not only relevant, but reliable."[1238]

Thus without any deliberate wrongdoing, the probative value of physical evidence in any given case can be cast in doubt by factors such as flawed procedures, incompetence, inattentiveness, and/or junk science.

The ease with which a forensic technician can be lulled into having unwarranted confidence in a procedure or his or her proficiency, and unwittingly violate "the first principle" of not fooling oneself, was unexpectedly demonstrated to the author while researching *U.S. v. Plaza*.[1239] A ruling in the *Plaza* case concerning the admissibility of fingerprint testimony incorporated testimony by Dr. Bruce Dudowle during a pretrial hearing in *U.S. v. Mitchell*.[1240] Dr. Dudowle, a geneticist in the FBI's Laboratory Division,[1241] described why methodological errors "can't" be made once a standardized protocol is set up:

> "When people spell words, they make mistakes. Some make consistent mistakes like separate, some people I'll say that I do this, I spell it S-E-P-E-R-A-T-E. That's a mistake. It is not a mistake of consequence, but it is a mistake. It should be A-R-A-T-E at the end.
>
> That would be an error. But now with the computer and Spell Check, if I set up a protocol, there is always Spell Check. *I can't make that error anymore.*"[1242]

The fundamental flaw in Dr. Dudowle's analysis was revealed as this article was written on several computers using the same popular word processing program installed on each computer.[1243] It is known that the word processing program's spell check feature works as expected on the primary computer that was used.[1244] Furthermore, a spell check was performed periodically as this was written, and no spelling errors were detected. However, when manual editing was begun, obvious spelling errors were encountered. So a spell check was performed and no errors were detected: Yet at that point there were *known* errors in the document. In the course of

120   Menace To The Innocent

trouble-shooting the anomaly, the document was loaded into a competing word processing program that could seamlessly read the document.[1245] That alternate program's spell-check feature correctly identified multiple spelling errors. Thus it was discovered that the word processing program on *one* of the computers used to write this affected the article's file in such a way that a spell check performed using that company's program *on any computer* returned a message that there were *no* spelling errors. Ironically, one of the words the program returned as spelled correctly was *seperate*: the very word Dr. Dudowle used to illustrate the perfection of an established spell-checking protocol.[1246] So the *belief* a protocol existed for the word processing program to correctly spell check the book you are reading allowed spelling errors to proliferate and remain undetected over a period of time. After a human actor observed that unintentional but very real errors existed in the text, they were corrected by the objective analysis of the text by a *seperate* – Oops! – *separate* word processing program produced by a different manufacturer that used an alternate methodology.[1247] The finding by the computer program used to write this book that misspelled words were correctly spelled is evidence Dr. Dudowle's testimony in *U.S. v. Mitchell* was erroneous. Methodological testing errors are not necessarily eliminated by a standardized protocol: Quite to the contrary, such a procedure can effectively obscure such errors from being readily apparent, and thus instill false confidence that there are no errors.[1248]

The spelling check episode also revealed a significant flaw in relying on the use of a double-blind technique by a *single* source to uncover testing errors: While it can eliminate bias in the evaluation of a sample (such as testing for spelling errors), a double-blind technique won't eliminate errors attributable to the use of a faulty methodology.[1249] Quite to the contrary, methodological errors by a single source can be masked under the guise of the test's alleged scientific basis: It is only by a second or third *independent* double-blind analysis that such errors can be expected to be detected and/or corrected.[1250] That process can be described as *triple-blind* testing.

Testing by multiple sources is consistent with the scientific methods requirement that conclusions be verifiable by repetition,[1251] and that a single adverse finding excludes it as a scientific determination.[1252] Indeed, as pointed out in the *McGraw-Hill Encyclopedia of Science and Technology*, the failure to use scientific methods "is a sure indicator of nonscience. In other words, a discipline where the scientific method plays no role is not a science. Thus, such fields as theology, literary criticism, psychoanalysis, homeopathy, graphology, and palmistry can hardly be regarded as scientific."[1253]

The magnitude of the unreliable expert prosecution testimony used to convict people in this country is hinted at by the fact that not a single federal or state court requires that testimony be grounded in methods consistent with the scientific method: *Daubert*[1254] and its progeny thus far have been used to conceal, rather than expose, the non-scientific nature of that testimony behind the façade of judicial officiousness.[1255]

It might seem somewhat unusual to a Martian silently observing the inner workings of human society, that prosecution favorable testimony by a lone "expert" can seal the fate of an accused person.[1256] However the blind faith placed in people considered to be experts at interpreting "the unseen, the indecipherable, or the incomprehensible,"[1257] is a phenomena of our time that extends to many areas of modern life.[1258] This overarching reliance on the explanations of presumed expert technologists was summed up by Jacques Ellul in *The Technological Society*: "technique has taken over all of man's activities."[1259]

The consequence of that blind faith in the law enforcement realm, is the power over a verdict held in the hands of the prosecution's expert witness(es)[1260] whose testimony may be fatally tainted by a reliance on one or more suspect techniques,[1261] or technician error.[1262] A more damaging consequence is when the mere appearance of a technique's reliability is used to substantiate testimony that the technique doesn't actually support.[1263] In both of those scenarios the truth of the suspect or non-existent probative value of the evidence may be known by the technician, others in the crime lab, and even the prosecutors, but it is concealed from the defendant, the judge, and the jurors.

This situation is exacerbated, and to a certain degree driven by the evaluation of alleged crime related evidence by a technician(s) with transparent ties to the investigative and prosecuting agencies that will benefit from having their narrative of the alleged crime reinforced.[1264] If for no other reason than human nature, it can safely be predicted that suspect expert testimony that conceals exculpatory or inconclusive test results, that exaggerates an evaluation's favorability to the prosecution, or that relies on manufactured, planted or non-existent physical evidence, will continue to contribute to wrongful convictions for at least as long as government run crime labs have a deciding role in evaluating the probative value of prosecution evidence.[1265]

That situation emphasizes the importance of recognizing that 'good enough for government work' is the leitmotif underlying the commonplace sub-standard performance of crime labs and their technicians.[1266] That is cause for concern because "'good enough for government work' ...means conduct and work that is third-rate, shoddy, and not worthy of praise."[1267] That same situation also prevails amongst freelance prosecution expert witnesses.[1268]

Furthermore, 'good enough for government work' dominates the performance of crime labs because internal monitoring and external oversight have not corrected the competency and reliability problems plaguing crime labs nationwide.[1269] A representative example of this is that although FBI officials proclaimed the agency was going to make meaningful changes in its crime lab operation after a critical report was issued by the Office of the Inspector General in April 1997[1270] – nothing substantive has changed.[1271] It was reported in *The FBI's Junk Science*, published *four years after* that report: "Far from being rectified, false testimony by FBI lab agents is still being presented in criminal trials around the country, influencing jurors and compromising trials to the point where it's difficult to determine the guilt or innocence of some defendants."[1272]

Three years after that, in February 2004, the National Research Counsel released a report, *Forensic Analysis: Weighing Bullet Lead Evidence*.[1273] That report concluded bullet lead tests the FBI lab has used since the 1960s to tie a bullet fragment from a crime scene with a bullet linked to a defendant, could only be considered as *circumstantial* evidence in conjunction with other unrelated supporting evidence.[1274] The NRC study found that contrary to the testimony of FBI lab technicians in countless cases: "The available data do not support any statement that a crime bullet came from, or is likely to have come from a particular box of ammunition, and references to "boxes" of ammunition in any form is seriously misleading under *Federal Rule of Evidence* 403."[1275] Although the study was conducted at the request of the FBI, which has used bullet fragment evidence in about 2,500 cases since the early 1980s, the agency rejected the NRC's conclusions.[1276]

Since the problems afflicting crime labs are systemically imbedded in the design of the techniques they employ, there is no reason to believe – even in a Pollyanna scenario – that they will diminish, much less be eliminated: Business as usual – dominated by the principle of 'good enough for government work' – will continue to flourish unabated in the nation's crime labs.[1277]

Yet contrary to what might be presumed at first glance, the elimination of technician bias stemming from an adoption of double-blind techniques would not rectify that situation. Their use *would* have the dual impacts of causing a quantum increase in the quality of testing performed by crime labs on prosecution evidence, which would in turn result in a significant reduction in the quantity of cases in which a technician would be able to truthfully testify as to the probative value of that evidence.[1278] However since those consequences would contribute to a weakening, if not the outright obliteration of the prosecution's narrative of the crime in an untold number of cases, there is no practical expectation the FBI or any crime lab will *voluntarily* adopt, much less faithfully implement, scientifically sound double-blind techniques.[1279] It is simply not in the self-interest of a crime lab or its "customers" – prosecutors and police agencies – to be constrained by the effects that would follow an improvement in the reliability of its testing of evidence, and a technician's resultant testimony about its actual, and possibly zero probative value.[1280] The ease of a crime lab's subtle but effective sabotaging of double-blind testing procedures was demonstrated by the previous explanation of a supposedly error proof spell-check protocol that returned false results by unapparently being altered beneath the scenes.[1281]

At first glance, a possible remedy to counteract the self-interest of crime labs and prosecutors to resist conscientiously implementing double-blind procedures, would be to require a technician to meticulously document each step of the evidence handling and testing process, and further require those records to be made available to the defense.[1282] However malevolent technicians and/or their supervisors could easily defeat that superficial safeguard by simply manufacturing fake records.[1283] Thus the possibly well intentioned attempt to increase the accountability of a crime lab's testing procedures, by increasing their documentation, could have the exact opposite effect of strengthening insubstantial testimony that is *now* unsupported by such documentation.[1284]

The dysfunctionality of crime labs is a fundamental feature of their design: Consequently, their operation can no more be substantively restructured to eliminate their pro-prosecution bias than a Zebra can will away its strips: They are both subject to the constraint expressed in Popeye's observation – "I Yam What I Yam!"[1285]

Consequently the only realistic assurance that insubstantial expert evidence isn't used against a defendant cloaked by the presumption of

Menace To The Innocent

innocence[1286] – and who may very well be actually innocent – is to approach the problem from an entirely fresh direction.

Methodic Doubt Explained

Fortunately there is a solution to that problem that with disarming ease would positively revolutionize the degree of certainty attached to expert evidence: The adoption of a principle of Cartesian philosophy as the guiding light in the quest to reasonably ensure the substantialness of expert evidence.[1287] That principle is known as *methodic doubt*, which is to search "for certainty by systematically though tentatively doubting everything."[1288] The reason *methodic doubt* works as a search for the truth is: "The hope is that, by eliminating all statements and types of knowledge the truth of which can be doubted in any way, one will find some indubitable certainties."[1289] *Random House Webster's Unabridged Dictionary* (1999) defines "indubitable" as, "that cannot be doubted; … unquestionable."

If there is any process requiring an "indubitable" certainty of the truthfulness of a statement of knowledge, it is a criminal prosecution that can result in a penalty of death or life imprisonment in the most severe cases.[1290] The achievement of an extraordinary level of certainty about the probative value of evidence by setting "aside as conceivably false all statements and types of knowledge that are not indubitably true," excludes the simplistic solution of merely having another laboratory review the testing process that was used to analyze the prosecution's evidence, or even having it examine the evidence.[1291] When that situation occurs today, and if the second expert (typically employed by the defense) disagrees with the testimony of the prosecution's expert, a "he said, she said" scenario exists that doesn't resolve the problem of achieving an "indubitable" certainty about what the truth is.[1292] Uncertainty about the evidence's probative value created by that conflict, is then imbedded in a case by judges and jurors who invariably resolve any doubts they may have in favor of the *opinion* expressed by the prosecution's expert.[1293] This pro-prosecution attitude is so powerful that it crosses national boundaries: John McManus, Project Coordinator of Miscarriages of Justice Scotland (MOJO Scotland), has observed: "… we are to quick to believe the views of prosecution experts based on expediency, tunnel vision or malfeasance…"[1294]

This deficiency in the current manner of evaluating what is offered as expert testimony underscores the lack of a scientific basis that all too often underlies it: So much so, that when there is a fundamental conclusatory disagreement between a prosecution expert and an expert for the defense concerning the probative value of the prosecution's evidence, a farcical courtroom situation exists that is more befitting of a nightclub comedy routine than a serious judicial inquiry.[1295] Yet in *Daubert* the Supreme Court rejected courtroom adversarial testing of evidence's probative value as an admissibility factor, and specifically endorsed the need to arrive at that conclusion by a "scientific methodology" that seeks to determine falsifiability through testing.[1296] Indeed, in *Daubert's* majority opinion Justice Blackmun observed: "Scientific methodology today is based on generating hypotheses and testing them to see if they can be falsified; indeed, this methodology is what distinguishes science from other fields of human inquiry."[1297] So the scientific validity of testimony is not in any way legitimately determined by the decision of a judge or the vote of jurors – it is determined by the methodology used outside the courtroom to arrive at the conclusions testified to in court.[1298]

The pervasive pro-prosecution bias of the current courtroom testing methodology can be neutralized if a cross-analysis of the evidence's evidentiary value is given equal weight in the eyes of the judge and the jury.[1299] *Methodic doubt* can accomplish that by eliminating the prevailing reliance on pathological prosecution expert testimony to determine an accused person's fate.[1300] That benefit is particularly important because the more doubtful a person's guilt, the more important insubstantial evidence is to the prosecution.[1301] Furthermore, the integrity of the judicial process would be bolstered by ascertaining the indubitable truthfulness of evidence and testimony that otherwise must be considered as suspect, and thus any conviction substantively relying on it is automatically tainted.[1302]

The way *methodic doubt* can seamlessly be integrated as a normal part of all criminal prosecutions involving expert testimony, would be for *the defense* and *the court*[1303] to each retain a different independent forensic laboratory or expert to evaluate the prosecution's allegedly incriminating evidence.[1304] It would be consistent with enforcing the prosecution's burden of proof[1305] for both of those laboratories to independently evaluate the alleged evidence and in their separate reports either confirm or dispute the prosecution's analysis of that evidence prior to its use against a defendant.[1306] It would likewise be consistent with enforcing the prosecution's proof burden, if when the three experts do not unanimously agree about the probative value of the evidence, it would be automatically deemed inconclusive. That would mimic what occurred in Britain until June 2001, when to avoid being automatically deemed as inconclusive, multiple examiners of fingerprint evidence had to independently agree the suspect's prints and the latent crime scene prints matched at a minimum of 16 points.[1307]

The justification for such a cut and dried declaration of inconclusiveness is built into *methodic doubts* requirement that indubitable certainty of a statement's truthfulness is dependent on the elimination of "all statements and types of knowledge the truth of which can be doubted in any way ..."[1308] Doubt about the scientific value of prosecution evidence can literally be defined as when one or more multiple experts independently from each other arrive at substantively different conclusions about the evidence. The fundamental soundness of this approach is that when the prosecution's evidence has substantial scientific merit, three experts independent of one another can no more be expected to disagree than if they were considering the sum of 2+2. If one of the experts arrives at a conclusion different than the others in analyzing evidence, then a "Houston we have a problem" scenario exists that inherently casts doubt on the evidence's probative value.[1309] Furthermore, the probability of corresponding erroneous test results *decreases* significantly as the number of independent analyses of evidence *increases*.[1310] A 20% probability of an erroneous analysis by one technician, for example, decreases by a factor of over 20,000 to a .008% probability that three analysts, independent of one another, will make the same error.[1311] Thus the veracity of expert testimony is verified by application of methodic doubt's principles to the analysis of prosecution evidence.

The conviction of innocent people by the prosecution's misrepresentation of expert testimony to judges and jurors as scientific is not limited to this country.[1312] It has resulted in the suggestion by John McManus of MOJO Scotland that prospective expert testimony should be analyzed by multiple experts to determine its scientific merit: "If it's a science, then the facts should be agreed before the trial proceeds; if they cannot come to agreement, then it should not be allowed in court, as the onus is on innocent until proven guilty."[1313] That suggestion is consistent with the principle of indubitable certainties underlying methodic doubt.[1314]

The straightforward answer to critics of such a cut and dried evaluation process is that if evidence the prosecution considers critical to its case is declared to be "inconclusive," then assuming the police agency involved conducted a thorough investigation, charges may simply have to be dropped in the interest of preserving both the appearance and the principle of justice.[1315] This is simply a recognition that in this country's judicial scheme an accused person's presumption of innocence is considered to legally trump the inability of the prosecution to build a case overcoming that protective shield.[1316]

Furthermore, the stated purpose of the Federal Rules of Evidence is "that the truth may be ascertained and proceedings justly determined."[1317] Justice Breyer noted in his concurring opinion in *General Electric vs. Joiner* (1997) that "Judges are not scientists."[1318] As a solution to the lack of professional training by judges to make science dependent decisions, Justice Breyer suggested that in accordance with *Federal Rules of Evidence* Rule 706: "[A] judge could better fulfill this gatekeeper function if he or she had help from scientists. Judges should be strongly encouraged to make greater use of their inherent authority … to appoint experts … . Reputable experts could be recommended to courts by established scientific organizations, such as the National Academy of Sciences or the American Association for the Advancement of Science."[1319]

However there is a fundamental problem with Justice Breyers' suggestion in *General Electric*: In and of itself it would do nothing to solve the favoring of prosecution expert testimony by jurors, or even judges for that matter.[1320] When a judge followed the recommendation of its appointed expert(s) to allow prosecution "expert" testimony, then the lack of agreement with one or more defense experts would result in exactly the same situation that exists today: The jury would invariably accept the prosecution expert's testimony as the most authoritative, and the judge would favor it since his or her own experts endorsed it.[1321] Justice Breyers' suggestion is nothing more than a timid gesture paying homage to the spirit of ascertaining the truthfulness of expert testimony – while it would in fact be meaningless at actually accomplishing that end.

In contrast, there would be a marked increase in the reliability of expert testimony resulting from adoption of the scientifically sound procedure of relying on a triad of experts independent from one another, that would not only be consistent with *methodic doubt*, but would have the side-effect of dramatically increasing the reliability of a criminal verdict involving expert testimony.[1322]

Chapter 19

Crime Labs Are A 20th Century Invention That Contribute To Shortshrifting Reasonable Doubt

The proposition that there should be indubitable certainty that expert testimony implicating a defendant is grounded in reality, is not only supported by the consistency of that approach with application of the scientific method to a criminal proceeding, but also by the history of crime labs in this country.[1323] The world's first lab dedicated to analyzing crime scene evidence was established in France in 1910.[1324] The first crime lab in the United States was established in Los Angeles in 1923.[1325] Six years later this countries second forensic lab was established in Evanston, Illinois.[1326] The FBI's crime laboratory was established in November 1932.[1327]

Yet it was noted in *Threads of Evidence* that it was decades after crime labs were established before police and prosecutors began seriously replacing "good old detecting" with forensic analysis.[1328] It was not until the 1960s that law enforcement began to seriously utilize forensics.[1329] In 1995 an FBI agent was quoted as describing this shift: "Gone are the days when we'd go to a crime scene and pick up whatever we could see. Nowadays we're more interested in evidence we can't see."[1330] In cases involving evidence invisible to judges, jurors, defense lawyers and the general public, an expert's interpretation is essential to an understanding of what that alleged evidence "we can't see" *is*, and what it *means*.[1331]

Prominent Victorian jurist Sir James Fitzjames Stephen's description of why police adopted techniques to extract confessions – "It is far pleasanter to sit comfortably in the shade rubbing red pepper into a poor devil's eyes than to go about in the sun hunting up evidence."[1332] – is apropos to why the use of evidence that can't be seen or understood without expert interpretation has been adopted with a vengeance by law enforcement: It enables a case file to be closed when "they can't find the hard facts that would make a tight case."[1333] Consequently, the interpretation of invisible evidence by prosecution associated expert witnesses gives law enforcement an almost free hand to secure an otherwise unobtainable conviction.[1334]

Thus applying a standard of indubitable certainty to expert testimony about evidence that can't be seen could be expected to have the same consequence as stopping police from taking the shortcut of manufacturing evidence by "rubbing red pepper" in a suspect's eyes: The police would have to "go about in the sun hunting up evidence" to overcome a suspect's presumed innocence instead of sitting "comfortably in the shade."[1335]

The fact there was no crime lab in this country for the first 136 years of this country's existence, and that the 20th Century was one-third over before the FBI's was established, is indicative that while they are not necessary for law enforcement to be effective, they do serve as an effective crutch aiding the short-circuiting of genuine police investigative work.[1336] To the degree that crime labs aid the successful masking over of a law enforcement investigation that was either incomplete or that did not find incriminating evidence against a suspect, the reasonable doubt that supposedly enshrouds every criminal defendant is shortshrifted, and the concept of justice subverted.[1337]

Yet there is no abiding reason for police agency (or other government funded) crime labs to even exist other than to provide a convenient "in-house" method of altering test results and/or slanting testimony to favor the prosecution's narrative of a case.[1338]

The numerous persistent problems known to plague crime labs indicates they are systemic, and not transitory in nature. Ten of those problems are:

- The absence of, or the failure to faithfully implement sound lab sterilization, and evidence handling and storage procedures to ensure there is no cross-contamination of evidence with lab debris, airborne substances, evidence from other crimes, or any other source.[1339]

- The absence of, or the failure to faithfully implement established testing protocols that are the underpinning of a scientific procedure.[1340]

- The failure to faithfully document evidence handling, test procedures, and the evidential basis of a conclusion.[1341]

- The failure to disclose that speculation about the probative value of evidence is testified to in scientific terms to make it appear substantive.[1342]

- The use of 'junk science' techniques to support test results and testimony.[1343]

- The failure to utilize corroborating test results by an independent third party scientist/lab to substantiate a test result and/or testimony.[1344]

- The failure to faithfully employ technicians with a scientific background *and* "real-world" simulated proficiency certification in the area of evidence analysis they are involved in.[1345]

- The failure to adequately police the testimony of lab technicians to prevent the embellishing of credentials.[1346]

- The failure to subject lab technicians to periodic double-blind proficiency testing to ascertain their skill level.[1347]

- The cultural and financial association of crime lab personnel with police agencies and prosecutors taints their subject judgments to be favorable to the prosecution.[1348]

The consequence of these pervasive deficiencies is that a typical high school science lab can be expected to function as more of a searcher of the truth than a typical crime lab, and the results of the former's experiments in a given case may be more reliable as evidence in court than the latter.[1349]

Thus ending the reliance of prosecutors on compliant police agency crime laboratories is important to increasing the scientific reliability of tests and expert testimony proffered by the prosecution.[1350] Consistent with ascertaining to an indubitable certainty the probative value of the prosecution's evidence, is a qualified independent forensic laboratory could randomly be selected to serve as the prosecution's analyst of the evidence in a particular case, and as an additional shield to preserve the integrity of the evidentiary process, that laboratory's personnel could be kept at somewhat of an arm's length from prosecution pressure to taint test results and testimony by being barred form having any direct communication with the police agencies and prosecutors involved.[1351] A court appointed intermediary could act as the communications go-between for all evidence to be tested and subsequent results and prospective testimony.

The dissolution of all federal and state law enforcement associated crime laboratories is an essential crucial step to ensure the evidentiary value of prosecution "scientific" evidence and expert testimony has a substantive and not an illusory foundation. [1352]

Chapter 20

Conclusion

State and federal criminal courts across this country are bastions of insubstantial, false and/or misleading expert prosecution evidence that is mistakenly believed by judges, jurors, defense lawyers, and society at large to be scientifically grounded.[1353] All too often suspect or discreditable testing procedures and/or theories underlies that insubstantial evidence.[1354] "Expert Syndrome" dominates the legal profession: alleged superior knowledge and training effectively elevates a person to a position of faith comparable to a witch doctor.

However, irrespective of the factors contributing to the prosecution's unreliable expert evidence, jurors, the trial judge, courtroom observers, and on review, appellate judges, can be sand-blind to the potentially exonerating value of the very physical evidence used to paint a defendant as guilty.[1355] The defendant and society at large are victims of the law enforcement system's pretense that it functions to impartially determine the fate of an accused person.[1356]

This situation is not new, since it has been the normal state of affairs since the latter half of the 20th century.[1357] That is when prosecutors began relying *en masse* on suspect scientific evidence to secure a conviction.[1358] Public exposure, however, will do nothing to effect meaningful reform: Since there is no feature in the current judicial framework to correct it,[1359] and as *de facto* agents of the prosecution with whom their self-interest is aligned, crime lab employees and outside experts have not, will not, and even more to the point, cannot be expected to do so.[1360]

The motive and opportunity for expert witnesses to support the prosecution's narrative of the crime is compounded by the knowledge that lying is so common by law enforcement associated personnel[1361] that it is the norm and not the exception.[1362] Furthermore, the pervasiveness of dishonest conduct is an accepted feature of the law enforcement process.[1363]

There are thus a number of compelling reasons why expert evidence is the mental drug of choice for prosecutors and police agencies to fix a judge, jurors and society at large with the high of believing it ties the defendant to crime related evidence. Consequently, any injection of rationally based certainty into the evaluation of the evidentiary value of expert testimony and/or allegedly scientific tests can only be accomplished without any dependence on, or interference by either the operational structure of the prosecutorial or police agency branches of the criminal prosecution system that benefit from insubstantial "scientific" evidence,[1364] or the courts that are prone to blithely allow it.[1365]

Under the current framework testimony by a prosecution expert about key evidence cannot be taken at face value.[1366] However the lack of incentive for crime labs to end their reliance on uncredible or short-shrifted techniques and the use of unproficient technicians, the lack of incentive for prosecutors to end their reliance on 'junk' expert testimony, and the lack of incentive for judges to exclude unscientific tests and testimony, can effectively be overcome by adopting the principle that indubitable certainty of the probative value of evidence must be established before it can be used against an accused person.[1367]

Outside and independent corroboration is necessary to determine what if any veracity prosecution expert testimony and any test(s) underlying it has.[1368] If that corroboration is not forthcoming, then the soundness of the testimony is automatically cast in doubt.[1369] This can unobtrusively be accomplished by adopting the pre-trial precaution of empowering separate experts acting as *independent* representatives of the defense and the Court, to analyze the prosecution's evidence and unilaterally veto the prosecution's use of expert testimony related to it.[1370]

A procedure intended to achieve a level of indubitable certainty about the scientific value of prosecution evidence and expert testimony is a minimal precaution to alleviate to the maximum degree possible, the damage inflicted on an innocent person from a wrongful prosecution, conviction and imprisonment.[1371] Particularly considering the deference accorded a prosecution expert witness' testimony concerning "invisible" evidence effectively usurps the fact finding function of the jury and/or the judge.[1372] The prosecution's paid expert witness(es) effectively functions as a "13th" juror, and is the single most important person in the courtroom determining a defendant's fate.[1373] Consequently, filtering evidence by application of "indubitable certainty" would have the further benefit of providing society at large with a valid reason to believe that the law enforcement system has an abiding concern with seeking to accurately distinguish the innocent from the guilty – and does not merely approach a prosecution as a sporting event to be won at all costs.[1374]

Intertwined with an effort to ensure the reliability of scientific tests and testimony, is that for as long as police agencies and prosecutors rely on crime laboratories culturally and financially tied to them, an inherent and unavoidable conflict of interest exists that taints every test result and every testimonial appearance of a crime lab technician.[1375] The only meaningful solution to solve that problem – and not merely put a band-aid over it – is to shutter every state and federal criminal evidence related crime laboratory. The sooner the better.

By removing prosecutors and judges from the decision loop of what evidence is presented in court as scientific, and ending the use of crime labs as a source of that evidence, the evidentiary value of scientific tests and testimony would be significantly more reliable.[1376] That would have the consequence of reducing the incidence of an innocent person's false implication in a crime.[1377] It would also have the auxiliary consequence of providing an effective answer to part of the question posed in the last sentence of *Tainting Evidence: Inside the Scandals at the FBI Crime Lab*: "...can forensic science in America ever be run by scientists?"[1378] Yes it can: But not until a cross-section of people independent of one another who are trained in rigorous scientific methods – and not lay judges, prosecutors, and unqualified and/or biased lab technicians or outside prosecution experts – have control over what is presented in court as scientific evidence and testimony.[1379] Only when the final say in the use of expert testimony and evidence is taken out of the hands of the police, prosecutors and judges involved in a particular case, will a person accused of a crime be protected from the use of insubstantial expert testimony and evidence to effectively frame the person for a crime he or she is innocent of committing. Only then can it legitimately be claimed that our society values, and doesn't just give meaningless lip service to the centuries old principle that a person is "innocent until proven guilty." This is no minor concern, because those four simple words when conscientiously applied by police, prosecutors and judges, provides each of us with a thin shield of protection from being wrongly accused, prosecuted and imprisoned – while when they are ignored, we are all vulnerable.

Works Cited

Books

Baden M.D., Michael and Marion Roach. *Dead Reckoning: The new science of catching killers*. Simon & Schuster, 2002.

Bedau, Hugo Adam, Michael L. Radelet and Constance E Putnam. *In Spite of Innocence*. Boston: Northeastern University Press, 1992.

Borchard, Edwin M. *Convicting the Innocent: Sixty-Five Actual Errors of Criminal Justice*. New Haven: Yale University Press, 1932.

Callahan, David. *The Cheating Culture: Why More Americans Are Doing Wrong to Get Ahead*. San Diego: Harcourt, 2004.

Coady, C. A. J. *Testimony: A Philosophical Study*. Oxford: Clarendon, 1992.

Cole, Simon A. *Suspect Identities: A History of Criminal Identification and Fingerprinting*. Cambridge, MA: Harvard UP, 2001.

Committee on Scientific Assessment of Bullet Lead Elemental Composition Comparison, National Research Council. *Forensic Analysis: Weighing Bullet Lead Evidence*. Washington, DC: The National Academies Press, 2004.

Connery, Donald S. *Convicting the Innocent: The Story of a Murder, a False Confession, and the Struggle to Free a "Wrong Man."* Brookline Books, 1996.

Ellul, Jacques. *The Technological Society*. New York: Random House, 1964.

Epstein, Sam. *The Riddle of the Stone Elephant*. New York: Grosset & Dunlap, 1949.

Faulds, Henry. *Guide to Finger-Print Identification*. Wood Mitchell (Publishers), 1905.

Fay, Stephen, Lewis Chester, and Magnus Linklater. *Hoax: the Inside Story of the Howard Hughes—Clifford Irving Affair*. New York: The Viking Press, 1972.

Freeman, Richard Austin. *The Red Thumb Mark*. London: House of Stratus 2001 (Originally published in the United Kingdom 1907).

Galton, Francis. *Finger Prints*. London: Macmillan and Co., 1892.

Harris, Richard. *The Fear of Crime*. New York: Praeger, 1969.

Hotorf M.D., Ken. *Uri-ine Trouble*. Vandalay Press, 1998.

Huff, C. Ronald, Arye Rattner, and Edward Sagarin. *Convicted But Innocent: Wrongful conviction and Public Policy*. SAGE, 1996.

Irving, Clifford. *Fake! the Story of Elmyr de Hory the Greatest Art Forger of Our Time*. New York: McGraw-Hill, 1969.

Irving, Clifford. *Hoax*. Danbury CT: Franklin Watts, 1981.

Kelly, John F. and Phillip K. Wearne. *Tainting Evidence: Inside the Scandals at the FBI Crime Lab*. New York: The Free Press, 1998.

Krawczak, M. and J. Schmidtke. *DNA Fingerprinting*. BIOS Scientific Publishers 1998.

Kurland, Michael. *How To Solve A Murder: The Forensic Handbook*. Macmillian, 1995.

Lee, Henry C. and R.E. Gaensslen, editors, *Advances in Fingerprint Technology*,

Second Edition. CRC Press, 2001.

Loftus, Elizabeth. *Eyewitness Testimony.* Cambridge: Harvard University Press, 1996.

National Research Council. *DNA Technology in Forensic Science.* Washington DC: The National Academies Press, 1992.

Noble, John Wesley, and Bernard Averbuch. *Never Plead Guilty; the Story of Jake Ehrlich.* New York: Farrar, Straus and Cudahy, 1955.

Park, Robert L. *Voodoo Science: The Road from Foolishness to Fraud.* New York: Oxford University Press, 2000.

Paulos, John Allen. *A Mathematician Reads the Newspaper:* Anchor, 1995.

Pfungst, Oskar. *Clever Hans (The Horse of Mr. Von Osten): A Contribution To Experimental Animal And Human Psychology.* New York: Henry Holt and Company, 1911.

Ratha, Nalini K., and Ruud Bolle. *Automatic Fingerprint Recognition Systems.* New York: Springer, 2004.

Rosenthal, Robert, and Lenore Jacobson. *Pygmalion in the Classroom Teacher Expectation and Pupils Intellectual Development.* New York: Holt, Rinehart & Winston, 1968.

Rosnow, Ralph L. and Robert Rosenthal. *Beginning Behavioral Research: A Conceptual Primer.* New York: Prentice Hall, 2001.

Scheck, Barry, Peter Neufeld and Jim Dwyer. *Actual Innocence.* New York: Penguin Putnam, 2001 rev. ed.

Shapiro M.D., Arthur K. and Elaine Shapiro, M.D. *The Powerful Placebo.* Baltimore: The John Hopkins University Press, 1997.

Silverstein, Herma. *Threads of Evidence.* Connecticut: Twenty-First Century Books, 1996.

Stengers, Isabelle. *The Invention of Modern Science* (Univ. of Minnesota Press 2000)

Swearington, M. Wesley. *FBI Secrets: An Agent's Expose.* Boston: South End Press, 1995.

Wehde, Albert and John Nicholas Beffel, *Finger-Prints Can Be Forged.* Chicago: The Tremonia Publishing Company, 1924.

Yant, Martin. *Presumed Guilty.* Amherst: Prometheus Books, 1991.

Newspaper, Magazine, and Online articles, and Television programs (Print and online)

Anderson, Curt. "Ex-FBI Lab Worker Guilty, Falsified DNA." *The Guardian* (London), May 19, 2004.

Applebome, Peter. "As Influence of Police Laboratories Grows, So Does Call for Higher Standards," *The New York Times*, December 22, 1987, p. 20A.

Ashenfelter, David. "Renowned Dog Handler Sentenced to 21 Months." *Detroit Free Press*, September 29, 2004.

Associated Press. "Cadaver Hunter Is Indicted: Woman is accused of planting evidence during victim searches." *The Detroit News*, August 21, 2003,

Metro/State section.

Associated Press. "Charges Dropped in Earprint Case." March 23, 2001 (http://www.truthinjustice.org/mccann.htm).

Associated Press. "FBI Reels From Wrongdoing by Workers." *St. Petersburg Times*, April 16, 2003, World/Nation section.

Associated Press. "Inmate Freed; Cleared in sex assault case after 15 years." *St. Augustine Record*, May 7, 2001.

Associated Press. "Michigan Cadaver Dog Handler Pleads Guilty to Federal Charges." *Lincoln Journal-Star*, March 11, 2004.

Audi, Tamara. "Bones of Contention: Cadaver-sniffing canine's finds are under suspicion; Handler of world-famous dog is charged with planting evidence." *Detroit Free Press*, July 14, 2003.

Berry, Steve. "Pointing a Finger at Prints." *Los Angeles Times*, Feb 26, 2002, at A1.

Bonner, Raymond. "Death Row Inmate Is Freed After DNA Test Clears Him." *The New York Times*, August 24, 2001, A11.

Boseley, Sarah. "Four People Dead is Four Too Many." *The Guardian* [London], August 6, 2001.

Bott, Rachelle. "Fred Zain: the retrial." *Sunday Gazette-Mail* (Charleston, WV), September 2, 2001, 1A.

Brand, David. "Fingerprint Evidence." *Cornell University News*, January 24, 2002.

Cannon, Angie. "Most Wanted: A Good FBI Lab." *The Oregonian* [Portland, OR], February 14, 1997, p. A22.

Colarossi, Anthony and Pamela J. Johnson. "Dad freed from life sentence in son's death." *Orlando Sentinel*, August 28, 2004.

Commoner, Barry. "Unraveling the DNA Myth: The Spurious Foundation of Genetic Engineering." *Harper's Magazine*, February 2002, pp. 39-47.

Cowans, Stephan. The Innocence Project, http://www.innocenceproject.org.

"CSI: Crime Scene Investigation – The Complete Second Season," *PopMatters Television Archive,* http://www.popmatters.com/tv/reviews/c/csi-season-2-dvd.shtml.

"Danny McNamee," *Innocent: Fighting Miscarriages of Justice*, http://www.innocent.org.uk/cases/dannymcnamee/index.html.

Daecher, Michael. "Forensic Fraud: No Motive, No Witness, and 99 Years in Jail: The Curious Conviction of Sonia Cacy." *Texas Observer*, August 28, 1998.

"Death of Lying Chemist Fred Zain." *Talk Left.com*. December 8, 2002. (Reprinted from, "Obituaries: Fred Zain, 52, Discredited W.Va. Police Chemist." *Newsday* [New York City], December 4, 2002).

Dent, Jackie. "Research Casts Doubt on 'Shaken Baby' Science." *The Guardian* [London], March 26, 2004.

Drennan, Kerry. "Tech Reviews Allegations Against Medical Examiner." *Amarillo Globe-News*, October 1, 2003.

Duffy, Shannon P. "Experts May No Longer Testify That Fingerprints 'Match'." *The Legal Intelligencer*, January 9, 2002.

Editorial. "The FBI's flawed lab," *The Oregonian*, February 16, 1997, B4.

Eggen, Dan. "Study Faults FBI Bullet Tests." *The Washington Post*, Feb. 11, 2004, p. A12.

Elliot, Janet. "Lawyers seek access to Bexar County Lab; New DA reviews Zain's work." *Texas Lawyer*, Aug. 2, 1999.

Ellis, Randy. "Forensic Chemist's Work Criticized." *The Oklahoman* [Oklahoma City],September 15, 1999.

"England to have a Criminal Court of Appeal," *The New York Times*, April 29, 1906.

Evans, Murray. "FBI Agent Sentenced in False Swearing Case." *The Enquirer* [Cincinnati, OH], June 18, 2003.

Evans, Murray. "FBI Scientist to Plead Guilty To Lying." *Miami Herald*, April 29, 2003.

"FBI Lab Investigation Widens." *TalkLeft.com*. April 28, 2003.

Feige, David. "The Dark Side of Innocence." *The New York Times*, June 15, 2003.

"Fingering Fingerprints." *The Economist*. December 14, 2000, Science and Technology section.

Fischer, Mary A. "The FBI's Junk Science." *Gentlemen's Quarterly*, January 2001, p. 115.

"Forensic Work Questioned." *Herald Sun* [Melbourne, AUS]. May 18, 2004.

Gibson, Ray. "Courted Expert Steps on Toes with Footprints." *Chicago Tribune*, April 6, 1986, p. 1.

Glover, Scott and Matt Lait. "Credibility of Hollywood private eye shattered." *The Seattle Times*, February 2, 2004, A7.

Glover, Scott and Matt Lait. "Even prosecutors sought pellicano for his expertise," *Los Angeles Times*, February 1, 2004.

Gould, Pamela. "2 Labs, 2 Cases, 2 Goofs." *The Free Lance-Star* [Fredericksburg], May 29, 2001.

Greenberg, Gary. "Is it Prozac? Or Placebo?" *Mother Jones*. Nov./Dec. 2003, p. 76 (6).

"Handwriting. Expert Has Found a Method of Copying Impressions," *Oakland Tribune*, September 13, 1913, p. 2.

Harden, Blaine. "FBI Faulted in Arrest of Oregon Lawyer." *Washington Post*, November 16, 2004, A02. (Report published in the *Journal of Forensic Identification*, November-December, 2004).

Heath, David. "FBI's Handling of Fingerprint Case Criticized." *The Seattle Times*, June 1, 2004.

Hey, Wilf. "Random Numbers." *PC Plus*, April 2004, at 176-178.

Jenkins, John A. "Experts' Day in Court." *New York Times Magazine*, December 11, 1983, p. 98.

Jha, Alok. "DNA Fingerprinting 'No Longer Foolproof'." *The Guardian* [London], September 9, 2004.

Johnston, David. "F.B.I. Removes 4 Lab Workers Including Critic of Crime Unit." *The New York Times*, January 28, 1997.

Joyce, Helen. "Beyond Reasonable Doubt." +*Plus magazine*, September 2002 (Issue 21), http://pass.maths.org.uk/issue21/features/clark.

"Judge Orders Retrial For Convicted Murderer." *Newstalk WJR 760am* [Detroit, MI], July 2004. At http://www.760wjr.com/Article.asp?PT=Local+News&id=33373.

Kershaw, Sarah. "Spain and U.S. at Odds on Mistaken Terror Arrest." *The New York Times*, June 5, 2004, National Section.

Khanna, Roma. "Credentials embellished: Transcript: Ex-lab chief told jury he had a Ph.D." *Houston Chronicle*, September 10, 2003.

Khanna, Roma and Steve McVicker. "New DNA test casts doubt on man's 1999 rape conviction." *Houston Chronicle*, March 10, 2003.

Khanna, Roma. "New Evidence Furor Hits HPD," *Houston Chronicle*, August 27, 2004, A1.

Khanna, Roma and Steve McVicker. "Private DNA lab can't replicate HPD results," *Houston Chronicle*, September 5, 2003.

Khanna, Roma and Steve McVicker. "29 cases missing evidence: 8 added to tally at police DNA lab," *Houston Chronicle*, November 12, 2003, p. 1.

Khanna, Roma and Steve McVicker. "Report outlines steps to validate crime lab," *Houston Chronicle*, August 7, 2003.

Kramer, Andrew. "Brandon Mayfield Opens Investigations Into His Arrest." *KATU 2 News* [Portland, OR], May 25, 2004.

Kresnak, Jack. "Bones-finder Says She Cheated." *Detroit Free Press*, March 12, 2004.

Kushner, David. "These Are Definitely Not Scully's Breasts: Inside One Man's Crusade To Save Gillian Anderson and the rest of the world from the plague of fake celebrity porn." *Wired Magazine*, November 2003, pp. 142-145.

Lander, Eric. "DNA Fingerprinting on Trial." *Nature*, 1989 Jun 15;339(6225):501-5.

"Lawyers Seek Access To Bexar County Lab: New DA Reviews Zain's Work." *Texas Law*, August 2, 1999.

Le Fanu, James and David Derbyshire. "In the Rush to Protect Children, 'Experts' Use Junk Science To Accuse Innocent Parents," *The Daily Telegraph* [London], December 13, 2003.

Lewis, Neil A. "House subcommittee rips into FBI director Louis Freeh." *The Oregonian*, March 6, 1997, p. A7.

Liptak, Adam. "Prosecutions Are a Focus in Houston DNA Scandal," *The New York Times*, June 9, 2003.

Liptak, Adam. "You Think DNA Evidence Is Foolproof. Think Again." *The New York Times*, March 16, 2003.

"Lossless Compression," *Webopedia* (http://www.pcwebopaedia.com)

"Lost, Mishandled Evidence May Mean New Appeals." *Florida Prison Legal Perspectives*, Vol. 6, Issue 6.

Martin, John P. "FBI Lab Again Scrutinized For Flawed DNA Tests." *New Jersey Star-Ledger*, April 8, 2003.

Main, Frank and Carlos Sandovi. "2 Charged in Roscetti Case." *Chicago Sun-Times*,

February 8, 2002.

McCloskey, James. "Convicting the Innocent." *Voice for the Defense*, Dec., 1989.

Mills, Steve, Maurice Possley and Kim Barker. "3 Roscetti inmates walk free: After 15 years, new world greets them as judge tosses convictions." *Chicago Tribune*, December 6, 2001, p. 1.

Mills, Steve and Maurice Possley. "Report alleges crime lab fraud: Scientist is accused of providing false testimony." *Chicago Tribune*, January 14, 2001.

Neergaard, Lauran. "FDA: Some antidepressants need stronger suicide warnings." *The Seattle Times*, March 23, 2004. at Main Section.

Newman, Andy. "Fingerprinting's Reliability Draws Growing Court Challenges." *The New York Times*, April 7, 2001.

Olsen, Lise. "Crime Lab Worker Failed To Qualify To Test Hair Samples." *Seattle Post-Intelligencer*, January 2, 2002.

Olsen, Lise and Roma Khanna. "DNA lab analysts unqualified: Review Finds Education, Training Lacking," *Houston Chronicle*, September 7, 2003.

Paris, Margaret L. and Julie Armstrong. "Who Can We Trust If Not the Police?" *Seattle Post-Intelligencer*, Oct. 24, 1995. p. A9.

Possley, Maurice. "Crime lab defends itself, analyst: City unit cited in allegations," *Chicago Tribune*, July 20, 2001.

Possley, Maurice. "Ex-Inmate Exonerated of Rapes Tries To Get His Life In Order." *Chicago Tribune*, June 29, 2000, A4.

Possley, Maurice and Steve Mills. "Governor pardons Roscetti 4: Action paves way for ex-inmates to get compensation from state fund," *Chicago Tribune*, October 17, 2002.

Possley, Maurice and Ken Armstrong. "Lab Tech in Botched Case Promoted: Testimony Helped Wrongfully Convict Man of Rape," *Chicago Tribune*, March 19, 1999, Metro Chicago Sec., p. 1.

Press Association. "Mother Cleared of Infant Sons' Deaths." *The Guardian* [London], December 10, 2003.

Press Release. "DOJ Aware of Problems in FBI's DNA Lab." *National Association of Criminal Defense Lawyers*, NACDL.org, November 25, 1997.

"Quincy M.E." *The Quincy Examiner*. http://www.inquire.net/quincy/index3.shtml.

Rippel, Amy C. and Anthony Colarossi. "Dead Baby's Dad Closer to Retrial." *Orlando Sentinel*, March 27, 2004.

Rippel, Amy C. and Henry Pierson Curtis. "Criticism, Blunders Mar Medical Examiner's Exit." *Orlando Sentinel*, August 31, 2003.

Shannon, Elaine. "The Gang That Couldn't Examine Straight." *Time Magazine*, April 18, 1997, pp. 30-31.

Shepardson, David. "Dog-handler Pleads Guilty." *The Detroit News*, March 12, 2004.

Sherrer, Hans. "Baby Alan's Mysterious Death: Alan Yurko is innocent, so who is responsible for his son's death?" *Forejustice.org*, Oct. 15, 2003.

Sherrer, Hans. "Introduction to The Innocents, Part One." *Justice Denied: the magazine for the wrongly convicted*, March 1999, Vol. 1, Issue 2.

Sherrer, Hans. "It's About Time! Fred Woodworth's Questioning of Fingerprint Evidence Is Long Overdue." *Justice Denied the magazine for the wrongly convicted*, Vol. 2, Issue 8, p. 10.

Sherrer, Hans. "Medell Banks Jrs.' Conviction for Killing A Non-Existent Child Is Thrown Out As A "Manifest Injustice"." *Justice Denied: the magazine for the wrongly convicted*, March 2003, Vol. 2, Issue 9, pp. 31-35.

Sherrer, Hans. "Prosecutors Are Masters at the Art of Framing People." *Justice Denied the magazine for the wrongly convicted*, Vol. 1, Issue 9, p. 22.

Sherrer, Hans. "Prosecutorial Lawlessness is its Real Name." *Justice Denied the magazine for the wrongly convicted*, Vol. 1, Issue 6, p. 16.

Sherrer, Hans. "Sally Clark's Conviction of Murdering Two of Her Children is Quashed After Discovery the Prosecution Concealed Evidence of Her Innocence." *Forejustice.org*, February 22, 2003 (http://forejustice.org/wc/sally_clark_freed.htm).

Sherrer, Hans. "The Innocents: the Prosecution, Conviction, and Imprisonment of the Innocent, Introduction (Part One)." *Justice Denied*, Vol. 1, Issue 2, p. 32.

Sherrer, Hans. "Landmark Study Shows the Unreliability of Capital Trial Verdicts." *Justice Denied*, Vol. 2, No. 2 (Nov. 2000).

Sherrer, Hans. "Twice Wrongly Convicted of Murder – Ray Krone Is Set Free After 10 Years." *Justice Denied: the magazine for the wrongly convicted*, March 2002, Vol. 2, Issue 8, pp. 25-26.

Singer, Christopher M. "Cadaver-dog Case May Boost Appeals." *The Detroit News*, August 22, 2003.

Singer, Christopher M. "Retrial Order to be appealed." *The Detroit News*, Aug. 15, 2004.

Solomon, John. "Conviction Overturned After Years In Prison: shoddy FBI lab work blamed for reversal in man's 1992 case." *Seattle Post-Intelligencer*, May 28, 2003, p. A3.

Solomon, John. "Panel Finds Flaws in FBI Bullet Analysis." *Charlotte Observer*, November 21, 2003.

Solomon, John. "Probe of FBI's DNA Lab Practices Widens." *FoxNews.com* (Associated Press), April 29, 2003., Available at, http://www.foxnews.com/story/2003/04/28/probe-fbi-dna-lab-practices-widen.html.

Staff. "Mother Cleared of Killing Babys." *BBC News*, June 11, 2003, at, http://news.bbc.co.uk/2/hi/uk_news/england/berkshire/2982148.stm.

Suro, Roberto and Pierre Thomas. "FBI lab errors taint numerous cases." *The Oregonian* [Portland], February 16, 1997, A23.

Tanner, Robert. "Standards, Autonomy Sought for Nation's Crime Labs." *The Seattle Times*, July 7, 2003, Nation section.

Teichroeb, Ruth. "Oversight of crime-lab staff has often been lax." *Seattle Post-Intelligencer*, July 23, 2004.

Teichroeb, Ruth. "Shadow of Doubt." *Seattle Post-Intelligencer*, March 13, 2004,

A1, A7.

Teichroeb, Ruth. "State Patrol fires crime lab scientist: His testimony in Montana cited; internal audit is downplayed." *Seattle Post-Intelligencer*, March 24, 2004, B1, B5.

Underwood, Graham. "Freedom nears for Erdmann as judge clears way for parole." *Lubbock Avalanche-Journal*, February 26, 1997.

Verkaik, Robert. "Up to 40 child-killing convictions in doubt." *The Independent* [London], September 12, 2004.

Vigoda, Ralph. "Fingerprints Put To Test." *Philadelphia Enquirer*, January 28, 2003.

Webb, Jaci and Greg Tuttle. "Free After 15 years." *Billings Gazette*, Oct 2, 2002.

Wecht, Cyril H. "DNA Testing: challenging the gold standard." *Tallahassee Democrat*, June 15, 2003, A1.

Weinstein, Henry. "Death Penalty Foes Mark a Milestone: Arizona convict freed on DNA tests is said to be the 100th condemned U.S. prisoner to be exonerated since executions resumed." *Los Angeles Times*, April 10, 2002.

Welles, Orson. *F for Fake* (1973, ASIN: 6303473261).

"What Jennifer Saw." *PBS Frontline*, http://www.pbs.org/wgbh/pages/frontline/shows/dna/cotton/compensate.html.

Wiley, John K. "State Patrol fires embattled crime-lab scientist." *The Seattle Times*, March 24, 2004, p. B3.

Wilkens, John. "Charges Are Dismissed In 1983 Death." *San Diego Union-Tribune*, September 4, 2004.

Woodworth, Fred. "A Printer Looks At Fingerprints," *Justice Denied*, Vol. 2, Issue 8, March 2002.

Woodworth, Fred. *The Mystery and Adventure Series Review*, No. 34, Summer 2001.

Woffinden, Bob. "Earprint Landed Innocent Man in Jail For Murder." *The Guardian* [London], January 23, 2004.

Yardley, Jim. "Inquiry Focuses on Scientist Used by Prosecutor." *The New York Times*, May 2, 2001.

Yurko, Francine. "Triumph Over Tragedy." *Justice Denied*, Issue 23, Winter 2004, 10-11.

Zaitz, Les and Noelle Crombie. "FBI Apologizes to Mayfield." *The Oregonian*, May 25, 2004.

Zaitz, Les. "FBI Case Against Oregon Lawyer Built on Blurry Fingerprint, Logic." *The Oregonian*, May 30, 2004.

Zaitz, Les. "Transcripts Detail Objections, Early Signs of Flaws." *The Oregonian*, May 26, 2004.

Journals, Reports, Scholarly Articles, and Research Sources

Acton, Lord. *Letter from Lord Acton to Bishop Mandell Creighton* (Apr. 3, 1887), *in* 1 *The Life and Letters of Mandell Creighton* ch. 13 (Louise Creighton ed. 1904).

Affidavit of Rickard K. Werder, May 6, 2004, *In Re: Federal Grand Jury* 03-01, No. 04-MC-9071 (USDC WD OR)

Allen, W. B. "The Laws of Science and the Rule of Law." A Presentation in the Forum "Science, Ethics and the Law" for the *Eighth Conference on Statistics, Science and Public Policy*, April 23-26, 2003 (http://williambarclayallen.com/presentations/Laws%20of%20science%20and%20the%20rule%20of%20law.htm).

Bernstein, David. "Junk Science in the United States and the Commonwealth." 21 *Yale J. Int'l L.* 123 (1996).

Brislawn, Chris. "The FBI Fingerprint Image Compression Standard." *Chris Brislawn's Website* (One of the designers of the FBI's national standard for wavelet-based compression (WSQ) of their fingerprint database.), http://www.disrv.unisa.it/~ads/corso-security/www/CORSO-9900/biometria/FBI.htm (last visited September 18, 2004).

Britannica, The Editors of Encyclopædia. "DNA Fingerprinting." *The New Encyclopædia Britannica*, Vol. 4, Rev. 15th Ed., 2002, pp. 140-141.

Britannica, The Editors of Encyclopædia. "Methodic Doubt." *The New Encyclopedia Britannica*, Vol. 8, Rev. 15th Ed., 2002, p. 70.

Burk, Dan L. "DNA Identification: Possibilities and Pitfalls Revisited." 31 *Jurimetrics* 53 (Fall 1990).

Carnell, Brian. "The Myth of Fingerprints." *Skepticism.net*, April 10, 2001.

Cohen, Ed. "The Placebo Disavowed: Or Unveiling the Bio-Medical Imagination." *The Yale Journal of Humanities in Medicine*, p. 6.

Connors, Edward, Thomas Lundregan, Neal Miller and Tom McEwen. "Convicted by Juries, Exonerated by Science: Case Studies in the Use of DNA Evidence to Establish Innocence After Trial." *National Institute of Justice*, 1996.

Cooley, Craig M. "Forgettable Science, Forensic Science, and Capital Punishment: Reforming the Forensic Science Community to Avert the Ultimate Injustice." 15 *Stan. L. & Pol'y Rev* 2, (2003).

DeFrances, Carol J. "Prosecutors in State Courts, 2001." *Bureau of Justice Statistics Bulletin*, NCJ 193441 (May 2002), quoted at 8, at http://www.ojp.usdoj.gov/bjs/pub/pdf/psc01.pdf (last visited Sept. 21, 2004).

Donohoe, Mark. "Evidence Based Medicine and Shaken Baby Syndrome. Part 1: literature review, 1966-1998." *Am. J. Forensic Med. Pathol.*, 2003;24:239-42.

Durose, Matthew R. and Patrick A. Langan, Ph.D. "State Court Sentencing of Convicted Felons, 2000 – Table 4.1 Estimated number of felong convictions in State courts." *Bureau of Jutsice Statistics*, U. S. Dept. of Justice, June 2003, NCJ 198822.

Editorial. "The Evidence Base For Shaken Baby Syndrome: We need to question the diagnostic criteria." *British Medical Journal*, Vol. 328, March 27, 2004, 719-20.

"Fingerprint Evidence." *Nolo Press*. (http://www.nolo.com/lawcenter).

Fisk, Margaret Cronin. *Lawyers Free Chicago Trio After Retesting of Lab Sample: Technician Had Testified in Earlier Bad Conviction*, NAT'L L.J., December 17, 2001, A6.

Foley, Michael Oliver. "Police Perjury: A Factorial Survey." *U.S. Dept of Justice, National Institute of Justice* (2000).

Giannelli, Paul C. "Junk Science: The Criminal Cases." 84 *J. Crim. L. & Criminology* 105 (1993).

Gifford, Sidra Lea. "Justice Expenditure and Employment in the United States, 1999," *Bureau of Justice Statistics*, NCJ 191746, February 2002.

Haber, Lyn and Ralph Haber. "Double-Blind Procedures in Forensic Identification and Verification," *Human Factors Consultants*, March 9, 2002, at 1-3, 13, unpublished article available at, http://humanfactorsconsultants.com/research.html.

Haber, Lyn and Ralph N. Haber. *Error Rates for Human Latent Fingerprint Examiners*, 2003, at 7. Unpublished manuscript available at, http://humanfactorsconsultants.com/research.html. (last visited Sept. 21, 2004).

Hagerman, Paul J. "DNA Typing in the Forensic Arena." 47(5) *Am. J. Hum. Genet.* 876-877, Nov. 1990.

Hansen, Mark. "Believe It Or Not." 79 *A.B.A.J.* 64 (June 1993).

Hansen, Mark. "Out of the Blue." 82 *A.B.A.J.* 50 (February 1996).

IAI. *Proceedings*. 1927.

Imwinkelried, Edward J. "Evidence Law Visits Jurassic Park: The Far-Reaching Implication of the Daubert Court's Recognition of the Uncertainty of the Scientific Enterprise." 81 *Iowa L. Rev.* 55, 1995.

In the Matter of an Investigation of the West Virginia State Police Crime Laboratory, Serology Division, 438 S.E. 2d 501, 190 W.Va. 321 (November 10, 1993).

"Innocents Database." *Forejustice.org*,

Jonakait, Randolph. "The Meaning of Daubert and What That Means for Forensic Science." 15 *Cardozo L. Rev.* 2103, 2117 (1994).

Kaye, D.H. and Jonathan Koehler. "Can Jurors Understand Probabilistic Evidence?," 154 *J. Royal Sts. Soc'y Series A* 75, 77 (1991).

Kennedy, Helena. "Sudden unexpected death in infancy: A multi-agency protocol for care and investigation." *The Royal College of Pathologists and the Royal College of Paediatrics and Child Health*, September 2004.

Koehler, J. "DNA Matches and Statistics." 76 *Judicature* 222 (1993).

Koehler, J. "Error and Exaggeration in the Presentation of DNA Evidence at Trial." 34 *Jerimetics* 21 (1993).

L. H. L. "The "Finger-Print" Case." *Michigan Law Review*, Vol. 10, No. 5 (March 1912) pp. 396-401.

Langan, Patrick A., and Jodi M. Brown. "Felony Sentences in the United States, 1994." *Bureau of Justice Statistics*, U. S. Dept. of Justice, Rev. Sept. 17, 1997, NCJ 165149.

Lantz, Patrick, S. Stanton, and C. Weaver. "Perimacular Retinal Folds From Childhood Head Trauma: Case report with critical appraisal of current literature," *British Medical Journal*, 2004; 328: 754-756.

Lefcourt, Gerald B. "Remarks of Gerald B. Lefcourt, President NACDL" (Regarding "The FBI Laboratory: An Investigation into Laboratory Practices and Alleged Misconduct in Explosives-Related and Other Cases (April 1997)"), *National*

Association of Criminal Defense Lawyers, November 25, 1997, http://www.nacdl.org/MEDIA/pr000097.htm.

Lempert, Richard. "After the DNA Wars:skirmishing with NRC II." *Jurimetrics* 37: 439-468 (1997).

Lempert, Richard. "Some Caveats Concerning DNA As Criminal Identification Evidence: With Thanks to the Reverend Bayes." 13 *Cardozo L.Rev.* 303, 1991.

Lewontin, R.C. and Daniel Hartl. "Population Genetics in Forensic DNA Typing," 254 *Science* 1745, 1991.

Liebman, James, Jeffrey Fagan and Valerie West. "A Broken System: Error Rates in Capital Cases and What Can be Done About Them." *The Justice Project*, June 12, 2000.

Luscombe, Richard. "Canine Sleuth Finds His Record Blighted," *The Observer* [London, UK], September 21, 2003.

McBay, A.J. "Drug-analysis technology-pitfalls and problems of drug testing." *Clinical Chemistry 1987*, Vol. 33, Number 11(B) 33B-40B.

McFarlane, Bruce A. "Convicting The Innocent – A Triple Failure of the System." *Canadian Criminal L*aw, August 2003, p. 57.

McManus, John. "Cot Deaths and the Adversarial Justice System," January 22, 2004, article sent to author from MOJUK's email service, http://www.mojuk.org.uk/.

McRoberts, Alan L. *"Fingerprints: What They Can & Cannot Do!"* SCAFO, Vol. 10 Issue 6 (June 1994), at 1-7, http://www.scafo.org/library/100601.html.

Moenssens, Andre A. "Admissibility of Fingerprint Evidence and Constitutional Objections to Fingerprinting Raised in Criminal and Civil Cases." *Chicago-Kent Law Review*, Vol. 40, Issue 2, 85(39), Oct. 1963.

Pankanti , Sharath, Salil Prabhakar and Anil K. Jain. "On The Individuality of Fingerprints," *IEEE Transactions on Pattern Analysis and Machine Intelligence*, Volume 24, Issue 8 (August 2002), article available at, http://biometrics.cse.msu.edu/prabhakar_indiv_pami.pdf.

Paris, Margaret L. "Lying to Ourselves." 76 *Or. L. Rev.* 817 (1997).

Paris, Margaret L. "Trust, Lies, and Interrogations." 3 *Va. J. Soc. Policy & Law* 3 (1995).

Press Release. "FTC Charges "Miss Cleo" Promoters with Deceptive Advertising, Billing and Collection Practices." *Federal Trade Commission*, February 14, 2002. (See at, http://www.ftc.gov/opa/2002/02/accessresource.htm)

Press Release. "UCI Professor Finds DNA Laboratory Errors That Sent Wrong Man to Prison." *Scientific Testimony: An Online Journal*, March 11, 2003.

"Pretrial Identification Procedures Lineups, Showups, Photographic Arrays." *Clark County Prosecutors, For Police Officers*, December 2000 Bulletin.

"Prosecutor's Fallacy." *Wikipedia.org.*

Random House Webster's Unabridged Dictionary (1999).

Roth, Nelson E. "The New York State Police Evidence Tampering Investigation: Confidential Report To The Honorable George Pataki Governor Of The State Of New York." Albany, NY: New York State Police, Jan. 20, 1997.

"Sally Clark: More About the Statistics and the Appeal." *Sally Clark website,* http://www.sallyclark.org.uk/AppealStats.html.

"Scientific Methods." *McGraw-Hill Encyclopedia of Science and Technology,* Vol. 16, 119-120 (1997).

Sherrer, Hans. *How Many Innocents Are There?* (Feb. 8, 2003) (Unpublished manuscript, on file with the author.).

Sherrer, Hans. "The Complicity of Judges in the Generation of Wrongful Convictions." 30 *N. Ky. L. Rev.* 539 (2003)

Sherrer, Hans. "The Illusion of Tax Evasion." pp. 102-122, in *The Ethics of Tax Evasion,* Robert W. McGee, ed. (The Dumont Institute for Public Policy Research , 1998).

Sherrer, Hans. "The Inhumanity of Government Bureaucracies." *Independent Review,* Vol. 5, No. 2, 249 (Fall 2000), at 258-260.

Sherrer, Hans. "The Innocents Database," *Forejustice.org.*

Starrs, James E. "The Ethical Obligation of the Forensic Scientist in the Criminal Justice System." *Journal of the Association of Official Analytical Chemists,* vol. 54, no. 4 (1971), pp. 1-12.

Starrs, James E. "Saltimbancos on the Loose? Fingerprint Experts Caught in a Whorl of Error." *The Scientific Sleuthing Newsletter,* Vol. 12, No. 2, Spring 1988, p. 1.

"The FBI Laboratory: An Investigation into Laboratory Practices and Alleged Misconduct in Explosives-Related and Other Cases." *U.S. Dept. of Justice, Office of the Inspector General* (April 1997),

"The Oxford English Dictionary." *Oxford English Dictionary Online* (http://dictionary.oed.com).

The Merriam-Websters Dictionary, 10th Collegiate Ed.

Thompson, William C. "Accepting Lower Standards, The National Research Council's second report on forensic DNA evidence." *Jurimetics* 1997; 37(4): 405-424.

Thompson, William C. "A Sociological Perspective on the Science of Forensic DNA Testing," 30 *U.C. Davis L. Rev.* 1113 (1997).

Thompson, William C., Franco Taroni, and Colin G.G. Aitken. "How the Probability of a False Positive Affects the Value of DNA Evidence," 48 *J. Forensic Sci.* 47, 48 (2003).

Underwood , Richard H. "Evaluating Scientific and Forensic Evidence," 24 *AMJTA* 149, 162 (Summer 2000).

Legal Cases & Evidence Rules

Arizona v. Youngblood, 488 U.S. 51 (1988).

State of Arizona v. Youngblood, 734 P.2d 592, 153 Ariz. 50 (1986).

People of California v. Marshall, No. BA-069796 (Sup Ct. LA County, Calif.)

People of California v. Soto, 48 Cal.App.4th 924, 30 Cal.App.4th 340, 34 Cal.App.4th 1588, 39 Cal.App.4th 757, 43 Cal.App.4th 1783, 35 Cal.Rptr.2d 846 (Cal.App. Dist.4 11/22/1994)

Daubert v. Merrell Dow Pharmaceuticals, 509 U.S. 579 (1993)

Emperor v. Sahdeo, (India) 3 Nagpur, L.Rep. 1 (1904)

Frye v. United States, 293 F. 1013 (D.C. Cir. 1923)

General Electric Co. v. Joiner, 522 U.S. 136 (1997)

Horstman v. Florida, 530 So.2d 368 (Fla.App.2 Dist 1988)

Illinois v. Gregory Fisher, 124 S.Ct. 1200 (2004)

Parker v. Rex, (Australia) 14 Commw. L.R. 681; 3 B.R.C. 68 (1912)

People of Illinois v. Jennings, 252 Ill. 534, 96 N.E. 1077 (1911)

In Re Castleton, 3 Crim. App. R. 74 (Cohen 1909)

In re Winship, 397 U.S. 358 (1970)

Jackson v. Florida, 511 So. 2d 1047 (Fla.App.2 Dist 1987)

Kumho Tire Co. v. Carmichael, 526 U.S. 135 (1999)

Miller v. Pate, 386 U.S. 1, 6-7 (1967)

Minnesota v Roger Sipe Caldwell, 322 N.W. 2d 574, 580 (1982)

Parker v. Rex, (Australia) 14 C.L.R. 681; 3 B.R.C. 68 (1912)

Ratzlaf v. U.S., 510 U.S. 135, 141 (1994)

Schlup v. Delo, 513 U.S. 298 (1995)

The Estates of Deborah Marie Tobin, et al. vs. SmithKline Beecham Pharmaceuticals, No. 00-CV-025D (DC WY 2000)

U.S. v. Addison, 498 F.2d 741, 744 (D.C. Cir.1974)

U.S. v Llera Plaza, No. 98-362-10 (E.D.Pa. 01/07/2002)

U. S. v. Llera Plaza, 188 F.Supp.2d 549 (E.D.Pa. 03/13/2002)

U.S. v. Mitchell, CR No. 96-407 (E.D. PA)

U.S. v Mitchell, 199 F.Supp 2d 262 (E.D. PA 2002)

U.S. v. Mitchell, 145 F.3d 572 (3rd Cir.1998)

U.S. v. Parks, CR-91-358-JS (C.D. CA)

Washington v. Kunze, 988 P.2d 977 (Wash.App.Div 2 1999)

18 U.S.C. §1622 - Subornation of perjury

Federal Grand Jury 03-01, No. 04-MC-9071 (USDC WD OR)

Fed. R. Crim. Proc. Rule 16

Fed. R. Evi. 102

Fed. R. Evi. 403

Fed. R. Evi. 404

Fed. R. Evi. 702

Fed. R. Evi. 706

Index

Index   149

Endnotes

[1] The first known forgery of a fingerprint by a law enforcement officer was in 1925. Simon A. Cole, *Suspect Identities: A History of Fingerprinting and Criminal Identification* (Cambridge, MA: Harvard University Press, 2001), 278. The remainder of this article recounts numerous irregularities involving numerous forensic techniques that continue up to the present time.
[2] A good summary of these practices is: Gerald B. Lefcourt, "Remarks of Gerald B. Lefcourt, President NACDL" (Regarding "The FBI Laboratory: An Investigation into Laboratory Practices and Alleged Misconduct in Explosives-Related and Other Cases (April 1997)"), *National Association of Criminal Defense Lawyers*, November 25, 1997, http://www.nacdl.org/MEDIA/pr000097.htm (last visited September 15, 2004) [herinafter, *Remarks of Gerald B. Lefcourt*]; See also, Editorial, "The FBI's flawed lab," *The Oregonian*, February 16, 1997, B4. (The FBI crime lab engages in "shoddy work, withholding of relevant evidence from defense attorneys and outright bias in favor of prosecutions."). The situation at the FBI's crime lab that are symptomatic of problems with police crime labs nationwide remains unchanged. See, *infra* Chapter 3.I: The FBI's Crime Lab.
[3] Many such cases of innocent people convicted are listed in the *"Innocents Database*, Forejustice.org, at, http://forejustice.org/innocentsdatabase.htm (last visited Sept. 21, 2004).
[4] See e.g., Richard H. Underwood, "Evaluating Scientific and Forensic Evidence," 24 *AMJTA* 149, 162 (Summer 2000), at 162. https://uknowledge.uky.edu/cgi/viewcontent.cgi?article=1261&context=law_facpub (last visited November 15, 2003).
[5] *Id.* (The special expertise of a person is what is considered to qualify him or her to evaluate scientific evidence.)
[6] Fred Woodworth, "A Printer Looks At Fingerprints," *Justice Denied*, Vol. 2, Issue 8 (March 2002), http://justicedenied.org/printerwoodworth.htm (last visited September 21, 2004).
[7] Underwood, *supra* note 4, at 173 fn 117, quoting, Professor C.A.J. Coady, *Testimony: A Philosophical Study* (Clarendon Press, Oxford, 1992), 277.
[8] See e.g., *U.S. v. Addison*, 498 F.2d 741, 744 (D.C. Cir.1974) (Refers to the "mystic infallibility" jurors ascribe to testimony represented as scientific.)
[9] Underwood, *supra* note 4, at 166 fn 82 (John Allen Paulos, *A Mathematician Reads the Newspaper* (Anchor 1995) (mathematician author referring to juror confusion with numbers); see also D.H. Kaye & Jonathan Koehler, "Can Jurors Understand Probabilistic Evidence?," 154 *J. Royal Sts. Soc'y Series A* 75, 77 (1991).
[10] See e.g., William C. Thompson, Franco Taroni, and Colin G.G. Aitken, "How the Probability of a False Positive Affects the Value of DNA Evidence," 48 *J. Forensic Sci.* 47, 48 (2003). (Timothy Durham was convicted of raping an 11-year-old girl in Oklahoma based on erroneous expert testimony that his DNA matched that of a rapist, even though 11 alibi witnesses testified that he was in another state at the time of the attack.)
[11] A website explains *CSI: Crime Scene Investigation* in the following way: The ... team of forensic scientists who work the graveyard shift for the Las Vegas Police Department Crime Lab ... piece together how the crime was committed. Using the latest in equipment and technology, they gather the tiniest bits of evidence left behind at a crime scene to help nab the bad guys." "CSI: Crime Scene Investigation – The Complete Second Season," *PopMatters Television Archive*, at http://www.popmatters.com/tv/reviews/c/csi-season-2-dvd.shtml. (last visited September 19, 2004). Quincy M.E. was a weekly series on national television from 1976 to 1983. It starred Jack Klugman as a crime fighting medical examiner who was able to solve crimes based on clues from dead bodies. A website devoted to Quincy M.E. is, *The Quincy Examiner*, at http://www.inquire.net/quincy/index3.shtml (last visited Sept. 19, 2004).
[12] Thompson, *supra* note 10, at 48.

[13] *Id.* Durham was released in 1997 after four years imprisonment.

[14] *Id.*

[15] *Id.* (See the account of Timothy Durham's case).

[16] The author must credit this insight to Fred Woodworth, which he explains in, Woodworth, *supra* note 6, at 48. ("To an increasing degree, evidence is an abstraction far removed from the normal experience of the human beings who comprise juries. Thus the jury system, like all other aspects of statism, has migrated toward even further authoritarianism -in this case toward the reliance on the unseen, the indecipherable, or the incomprehensible, as delivered as fact by "expert witnesses.") *Id.*

[17] For example, in federal court, the testimony of experts about scientific evidence is governed by *Federal Rule of Evidence*, Rule 702. Testimony by Experts [herinafter, *Fed. R. Evi.* 702]: "If scientific, technical, or other specialized knowledge will assist the trier of fact to understand the evidence or to determine a fact in issue, a witness qualified as an expert by knowledge, skill, experience, training, or education, may testify thereto in the form of an opinion or otherwise, if (1) the testimony is based upon sufficient facts or data, (2) the testimony is the product of reliable principles and methods, and (3) the witness has applied the principles and methods reliably to the facts of the case."

[18] If the defense is financially able to do so, it will have their own expert(s) testify about the probative value of the prosecution evidence in dispute. This is as true in civil cases as criminal cases. The Supreme Court's 1993 *Daubert* case that replaced the *Frye* standard of determining the admissibility of scientific testimony in federal court under *Fed. R. Evi.* 702 was a civil, not a criminal case. See, *Daubert v. Merrell Dow Pharmaceuticals,* 509 U.S. 579 (1993). *Daubert* established a four pronged test for a federal court to use in considering the reliability of otherwise relevant expert testimony considered for admission under *Fed. R. Evi.* 702: four factors–testing, peer review, error rates, and "acceptability" in the relevant scientific community–which might prove helpful in determining the reliability of a particular scientific theory or technique, *Id.,* at 593-594.

[19] See e.g., the following references in, Underwood, *supra* note 4, at 152 ("... forensic scientists may be over-zealous or manipulated by their prosecutorial "customers," and that some "experts" may be down-right dishonest,..."); *Id.* at 167 ("It is no secret that expert witnesses can be "co-opted" by the prosecution – they may be little more than hired guns of the state."); at 168 ("Sometimes prosecutors will "shop around" until they find an expert who will tell them what they want to hear."); *Id.* at 175 (relating prosecutor's solicitation of bogus evidence).

[20] Peter Applebome, "As Influence of Police Laboratories Grows, So Does Call for Higher Standards," *The New York Times,* December 22, 1987, 20A, cited in, Martin Yant, *Presumed Guilty* (Prometheus Books 1991), 68. (Even if independent testing isn't conducted, a defendant may present expert testimony challenging the prosecution's portrayal of its evidentiary value to support a conviction.)

[21] *Id.* (Lab technicians and administrators, "often work hand in hand with prosecutors, while defense attorneys seldom have the resources to do their own forensic work. ") *Id.*

[22] In an Oregon case the author has knowledge of, a woman was convicted of the first degree murder of her husband with a handgun, during what she claims was a suicide attempt while under the influence of the prescription drug Paxil. The prosecution claims two vials of blood drawn from her at the time of her arrest were consumed by the crime labs standard tests, and that no blood remains for her to conduct an independent test for the presence of Paxil in her bloodstream at the time of the incident. The prosecution claims the crime lab did not test for the presence of Paxil, in spite of her assertion from the time of her arrest that she was under the influence of the drug at the time of the alleged crime. One of Paxil's known side effects is to induce suicidal thoughts and behavior in a percentage of its users who had no prior history

of engaging in them. See e.g., Lauran Neergaard, *FDA: Some antidepressants need stronger suicide warnings, The Seattle Times*, March 23, 2004. at Main Section. The most prominent case of a person reacting violently after taking Paxil is the Wyoming case of Donald Schell, who two days after beginning to take the drug, shot and killed his wife, his daughter and his granddaughter, before committing suicide. The family's survivors sued Paxil's manufacturer and were awarded a total of $6.4 million by a federal court jury in June 2001. See, *The Estates of Deborah Marie Tobin, et al. vs. SmithKline Beecham Pharmaceuticals*, Civil No. 00-CV-025D (DC WY 2000). See also, Sarah Boseley, "Four People Dead is Four Too Many," *The Guardian*, August 6, 2001, *at* http://www.guardian.co.uk/print/0,3858,4236013-103409,00.html (last visited September 21, 2004).

[23] See e.g., Michael Kurland, *How To Solve A Murder: The Forensic Handbook*, (Macmillian 1995), 5 ("As for training, many employees of this country's forensic laboratories are inadequately, incompletely, or improperly trained."). *Id.* See also, Applebome, *supra* note 20. ("...the work of the 3,500 or so forensic scientists in police crime labs is plagued by uneven training, a lack of certification and professional standards, and questionable objectivity.") *Id.*.

[24] See e.g., "Scientific Methods," *McGraw-Hill Encyclopedia of Science & Technology* Vol. 16, 119-120 (1997) (Among the aspects of the scientific method are: "testable consequences;" "repetition of the test;" and "reliability and accuracy."). *Id.*

[25] See e.g., Hugo Adam Bedau, Michael L. Radelet and Constance E Putnam, *In Spite of Innocence* (Northeastern University Press 1992), 141-152 (Relating that false expert testimony convinced a jury that red paint on a garmet that in fact didn't belong belong to Lloyd Miller, was the blood of a murdered young girl, and relying on those lies the jury convicted Miller and he was sentenced to death.) *Id.* See also, John F. Kelly and Phillip K. Wearne, *Tainting Evidence: Inside the Scandals at the FBI Crime Lab* (The Free Press 1998), 14-16. The authors write in regards to Thomas Curran, an FBI Special Agent assigned as an examiner in the agency crime lab's serology unit: "Tom Curran turned out to have lied repeatedly under oath about his credentials, and his reports were persistently deceptive, yet no one – FBI lab management, defense lawyers, judges – had noticed. When they did, there was no prosecution for perjury." *Id.* at 14.

[26] *Id.* at 16.

[27] An example is Lloyd Miller's conviction for murdering a girl based on the prosecutor's presentation of scientific testimony that the red substance on allegedly incriminating underwear was the victim's blood, when it was actually red paint. Bedau, *supra* note 25, at 141-152.

[28] *Id.*

[29] *Id.*

[30] *Id.*

[31] *Id.*

[32] *Id.*

[33] *Id.*

[34] *Miller v. Pate*, 386 U.S. 1, 6-7 (1967). For an account of the case, see, Bedau, *supra* note 25. See also, C. Ronald Huff, Arye Rattner, and Edward Sagarin, *Convicted But Innocent: Wrongful conviction and Public Policy* (SAGE 1996), 72. An interesting twist on Miller's case is that the Illinois Bar Association refused to discipline the prosecutor for his role in framing him for the girl's murder and sending him to death row. For examples of the many suspect tactics employed by prosecutors to secure a conviction, see, Hans Sherrer, "Prosecutors Are Master Framers*," Justice Denied*, Vol. 1, Issue 9; see also, Hans Sherrer, "Prosecutorial Lawlessness is its Real Name," *Justice Denied*, Vol. 1, Issue 6.

[35] Bedau, *supra* note 25, at 141-152.

[36] *Id.*

[37] The Innocents Database, *supra* note 3, has summaries of a number of cases involving innocent people wrongly convicted by a reliance of juries and/or judges on evidence erroneously considered to be scientific.

[38] An indicator of the increased use of DNA evidence is that it was reported by the Bureau of Justice Statistics, "In 2001 two-thirds of prosecutor offices reported the use of DNA evidence during plea negotiations or felony trials. This is an increase from 1996 when about half offices reported using DNA evidence during plea negotiations or felony trials..." Carol J. DeFrances, "Prosecutors in State Courts, 2001," *Bureau of Justice Statistics Bulletin*, NCJ 193441 (May 2002), quoted at 8, at http://www.ojp.usdoj.gov/bjs/pub/pdf/psc01.pdf (last visited Sept. 21, 2004).

[39] John Solomon, "Conviction Overturned After Years In Prison: shoddy FBI lab work blamed for reversal in man's 1992 case," *Seattle Post-Intelligencer*, May 28, 2003, at A3.

[40] Solomon, *supra* at A3.

[41] See e.g., C.A.J. Coady, *supra* note 7, at 289. Cited in Underwood, *supra* note 4, at 152 note 13.

[42] C.A.J. Coady, *supra* note 7, at 300. Cited in Underwood, *supra* note 4, at 152 note 14.

[43] The converse is true if a defense expert is successful at supplanting the jury's fact finding function, but it happens much less often than the reverse because of the general disparity of resources available to the prosecution as contrasted with the defense – which is typically handled by a resource challenged public defender agency.

[44] Insofar as this concern relates to DNA, considered the gold standard of physical evidence, Cyril H. Wecht, past president of the American Academy of Forensic Sciences wrote in June 2003, "There can be little doubt in the minds of trained, experienced forensic scientists that testing defects, backlog pressures, inadequately qualified personnel, and prosecutorial bias exist in many other DNA labs even though they have not yet been uncovered and publicly reported." Cyril H. Wecht, "DNA Testing: challenging the gold standard," *Tallahassee Democrat*, June 15, 2003, at A1, http://www.law-forensic.com/cfr_gen_art_13.htm (last visited September 21, 2004).

[45] Yant, *supra* note 20, at 66 (Quotes by Grunbaum from a 1985 symposium on forensic chemistry).

[46] Kurland, *supra* note 23, at 5. For the situation today, see e.g., Michael Baden, M.D. and Marion Roach, *Dead Reckoning: The new science of catching killers* (Simon & Schuster 2002), 232-234.

[47] Yant, *supra* note 20, at 66 (Quotes by Grunbaum from a 1985 symposium on forensic chemistry).

[48] *Id.*

[49] Kelly, *supra* note 25, at 30.

[50] *Id.* at 30.

[51] John A. Jenkins, "Experts' Day in Court," *New York Times Magazine*, December 11, 1983, at 103. (The technician errors included false positives and false negatives.) See also, Kelly, *supra* note 25, at 29-30. This examination was conducted over the four year period of 1974 to 1977. The prevalence of technician errors in crime lab testing continues. As explained elsewhere in the text, a 1995 proficiency test of 156 fingerprint examiners from across the country resulted in a 22% error rate. See e.g., *Id.* at 32. Proficiency tests in 1996, 1997 and 1998 had false positive rates of up to 15%. Cole, *supra* note 1, at 297.

[52] Jenkins, *supra* note 51, at 103. See also, Kelly, *supra* note 25, at 29-30. Although this examination was conducted over the four year period of 1974 to 1977, the prevalence of erroneous crime lab testing continues. See *Infra* notes 57-89 and accompanying text.

[53] *Id.*

[54] See, *infra* note 79 and accompanying text.

Menace To The Innocent

[55] The National Academy of Sciences has concluded that voiceprint theory has not been validated. See, Kelly, *supra* note 25, at 18.

[56] Voiceprint analysis isn't the only suspect testimony allowed in court. For example, *Daubert* challenges to fingerprint testimony examinations have thus far been unsuccessful, seemingly due to judicial bias to maintain the status quo, rather than any scientific basis for its admissibility being substantiated by the prosecution. See e.g. *infra*, Chapter 6.VIII: *The Mitchell Case (1999)*, and Chapter 6.IX: *The Plaza Case (2002)* (both discuss cases concerned with challenges to the admissibility of fingerprint evidence under *Fed R. Evi.* 702.).

[57] Ken Hotorf, M.D., *Uri-ine Trouble* (Vandalay Press 1998).

[58] *Id.* at 69, supported by fn 10, Mcbay, "AJ. Drug-analysis technology-pitfalls and problems of drug testing," *Clinical Chemistry* 1987, Vol. 33, Number 11(B) 33B-40B. When the labs knew the sample was a part of the test, they were able to detect the correct sample 95% of the time: Which means they were still *wrong* 1 out of 20 times. *Id.* This test showed the importance of blind proficiency testing to find the actual performance of a laboratories technicians, because the error rate increased by a factor of almost *1100%* when the laboratories did not know the sample was a part of the test – or a real-life sample. *Id.* The author, Dr. Holtorf also emphasized that to simulate real-world conditions, cross-reacting substances need to be involved in a blind test, just as they are in every real world sample tested for the presence of one or more drugs. *Id.*, at 69.

[59] Randolph Jonakait, "The Meaning of Daubert and What That Means for Forensic Science," 15 *Cardozo L. Rev.* 2103, 2117 (1994), referenced in Underwood, *supra* note 4, at 163-4. note 72.

[60] See, Lyn Haber and Ralph N. Haber, *Error Rates for Human Latent Fingerprint Examiners*, 2003, at 7. Unpublished manuscript available at, http://humanfactorsconsultants.com/research.html. (last visited September 21, 2004). Article scheduled to be published in, Nalini Ratha and R. Bolle, editors, "Advances in Automatic Fingerprint Recognition," *Springer-Verlag Publishers* (New York 2004).

[61] See, Haber, *supra* note 60 at 7.

[62] *Id.*, quote at 7.

[63] *Id.* at 7-8.

[64] *Id.* at 7-8. In 1983 for example, the average laboratory consensus erroneous identification rate was 9%. If two laboratory personnel collaborated on the test, the individual erroneous identification rate was 30% (.30 x .30 = .09). The average concensus percentage from 1983 to 1991 was 2.5%, so the average individual erroneous identification rate was 16% (.16*.16=.0256). *Id.* at 8.

[65] *Id.* at 7-8. In 1983 for example, the average laboratory consensus error rate was 24%. If two laboratory personnel collaborated on the tests, the individual error rate was 49% (.49 x .49 = .24). *Id.* at 7-8.

[66] *Id.* at 9.

[67] *Id.* at 9.

[68] *Id.* at 9.

[69] *Id.* at 9. Summaries of those test results are in CTS publications 9508, 9608, 9708, 9908, 0008, 0108, 01-516 and 01-517. *Id.*

[70] *Id.* at 9-10.

[71] *Id.* at 9-10. This is based on 67% of the responses by individuals, and 33% of the responses were by a consensus of two people in a laboratory. If an average of more than two people participated in the response of labs, then the percentages of erroneous individual responses would be higher than stated in the text.

[72] *Id.* at 9-10. This error rate includes false positives and false negatives. The high was 84% in 1996, and the low was 9% in 2000. The actual test results are reported for this, without an

adjustment for individual errors in the consensus crime lab responses, because the summary of errors is reported differently for the tests from 1983-1991 and 1995-2001.

[73] Cole, *supra* note 1, at 281. See also, Kelly, *supra* note 25, at 32.

[74] Kelly, *supra* note 25, at 32.

[75] Haber, *supra* note 60 at 9.

[76] Id. at 9.

[77] Id. at 9.

[78] Id. at 10. ("If these comparisons had been presented in court as testimony, one in five latent print examiners would have provided damning evidence against the wrong person. If these are consensus reports, more than half would have testified erroneously."). *Id.*

[79] For the ASCLD test results see, *supra* notes 64-65 and accompanying text; and for the CTS test results see, *supra* notes 71-77 and accompanying text.

[80] Haber, *supra* note 60, quote at 10.

[81] *Id.* at 9.

[82] *Id.* at 7-10.

[83] *Id.* at 9. ("Research on double blind testing procedures … show that test score results are inflated when people know they are being tested, and when they and their supervisors know of the importance of the test results.").

[84] *Id.* at 9. ("The … consensus error rate for erroneous identifications underestimates the error rate that actually occur in court."). *Id.*

[85] *Id.* at 16.

[86] *Id.* at 16.

[87] *Id.* at 16.

[88] *Id.* at 16. (emphasis added to original).

[89] (50*0%=0) + (50*21%=10.5) = 10.5/100 =10.5%. (50*10% =5) + (50*30%=15) = 20/100=20%.

[90] Lefcourt, *supra* note 2. In the early 1990s the FBI also conducted what was a de facto DNA proficiency test, when DNA samples were taken from 225 FBI agents 14 months apart. Using standard testing techniques and protocols, there was a 12-1/2% false positive rate – one out of eight DNA samples was matched to the wrong person - when the two sets of samples were matched by FBI crime lab technicians. Kelly, *supra* note 25, at 250.

[91] *Id.* (In contrast with a blind test, the technicians had the advantage of knowing they were being tested, and they were aware of the parameters of the test.).

[92] *Id.*

[93] *Id.*

[94] *Id.* (The test was supposedly readministered, with *all* the technicans passing. However that claim is suspect because the FBI has not produced any records concerning the alleged retest and its results. This scenario is consistent with the FBI's resubmittal of tests with the "correct" answers marked with "red dots," to the nine labs that did not initially match the suspect's prints with a crime scene print in 1999s *Mitchell* test. See, *United States v. Llera Plaza*, No. 98-362-10 (E.D.Pa. 01/07/2002); 2002.EPA.0000003 ¶ 268 note 23 <http://www.versuslaw.com> (last visited Sept. 21, 2004) (Cite as 179 F.Supp 2d 492 withdrawn when *U.S. v. Llera Plaza*, 188 F.Supp 2d 549 (E.D.Pa. 2002) was issued upon reconsideration.).

[95] *Id.* ("…all DNA/serology examiners in the DNA Unit (except possibly one) failed an open proficiency serology test in 1989."). *Id.*

[96] Kelly, supra note 25, at 15.

[97] *Id.* at 15.

[98] *Id.* at 15.

[99] *Id.* at 14. (Describes the pervasive perjury concerning test results by FBI lab technician,

special agent Thomas Curran.)

[100] *Id.* at 14.

[101] *Id.* at 14.

[102] *Id.* at 14.

[103] *Id.* at 16. (That sentiment was voiced by former FBI Director Louis Freeh.)

[104] For the active role prosecutors take in creating the circumstances underlying a conviction, see, Sherrer, *supra* note 34; see also, *Prosecutorial Lawlessness, supra* note 34.

[105] Kelly, supra note 25, at 14. More than a quarter century after Curran's activities were exposed in February 1975, the FBI crime lab is still not subject to quality controls such as blind proficiency testing by outside investigators, that are required of clinical lab technicians. See e.g., Thompson, *supra* note 10, at 53.

[106] Roberto Suro and Pierre Thomas, "FBI lab errors taint numerous cases," *The Oregonian*, February 16, 1997, at A23.

[107] *Id.*

[108] Elaine Shannon, "The Gang That Couldn't Examine Straight," *Time Magazine*, April 18, 1997, at 30-31.

[109] Kelly, supra note 25, at 37-38.

[110] Lefcourt, *supra* note 2. ("…there was a DNA lab technician or examiner – to quote from a DOJ document released today, – "who would determine if suspects were Afro-Americans. If so, he would manipulate test results to prove guilt.""). *Id.*

[111] David Johnston, "F.B.I. Removes 4 Lab Workers Including Critic of Crime Unit," *The New York Times*, January 28, 1997.

[112] *Id.*

[113] Angie Cannon, "Most Wanted: A Good FBI Lab," *The Oregonian*, February 14, 1997, at A22. The FBI's attempts to silence Frederic Whitehurst resulted in a settlement to him of 1.16 million contingent on his resignation. The FBI's way of dealing with the problem of operating a substandard crime lab was to get rid of the one man willing to reveal its deficient and unethical practices to the public.

[114] Johnston, *supra* note 111. The FBI has continued J. Edgar Hoover's policy of not tolerating "… any whistleblowers in the FBI." M. Wesley Swearington, *FBI Secrets: An Agent's Expose* (South End Press 1995), 55. See also, Hans Sherrer, "The Inhumanity of Government Bureaucracies," *Independent Review*, Vol. 5, No. 2, 249 (Fall 2000), at 258-260.

[115] See e.g., Neil A. Lewis, "House subcommittee rips into FBI director Louis Freeh," *The Oregonian*, March 6, 1997, at A7. In just one 1991 case that was cited in a study by the Justice Department's inspector general, " ... a lab witness overstated test results ... [And] In addition to overstated testimony in Vanpac [the case], the report found the lab lacked databases to support its conclusions, used unvalidated tests, lacked written test procedures, inadequately documented why it discounted test results that undercut its conclusions and lacked any record for some test." This litany of actions resulted in the conviction of what is perhaps an innocent man. *Id.*

[116] Kelly, *supra* note 25, at 37-38.

[117] *Id.* at 37-38.

[118] *Id.* at 37-43.

[119] *Id.* at 37-43.

[120] *Id.* at 42.

[121] "The FBI Laboratory: An Investigation into Laboratory Practices and Alleged Misconduct in Explosives-Related and Other Cases," *U.S. Dept. of Justice, Office of the Inspector General* (April 1997), at 30-31. Available at, http://www.usdoj.gov/oig/special/9704a/index.htm (last visited September 15, 2004).

[122] Kelly, *supra* note 25, at 58-63. (Relates a full history of Frederic Whitehurst's complaints

to his superiors in the FBI concerning crime lab deficiencies.)

[123] *Id.* at 2, 37-43, 59-60..

[124] Lefcourt, *supra* note 2.

[125] *Id.*

[126] *Id*

[127] *Id*

[128] Kelly, *supra* note 25, at 2, 60.

[129] *Id.* at 2, 60.

[130] *Id.* at 2 -3. (The investigation only included three of seven units in the lab's Scientific Analysis Section, and none of the lab's other 20 units. (See also, Lefcourt, *supra* note 2.) The investigated units were the Explosive Unit (EU), the Chemistry – Toxicology Unit (CTU) and the Materials Analysis Unit (MAU) – but not the DNA Unit or Fingerprint Analysis Unit. See, The FBI Laboratory, *supra* note 121, at Part One: Executive Summary.

[131] For separate news stories related to these three situations see, Murray Evans, "FBI Scientist to Plead Guilty To Lying," *Miami Herald*, April 29, 2003, at http://www.miami.com/mld/miamiherald/news/politics/5744064.htm?template=contentModul es/printstory.jsp (last visited Sept. 21, 2004); John P. Martin, "FBI Lab Again Scrutinized For Flawed DNA Tests," *New Jersey Star-Ledger*, April 8, 2003, available at, http://www.whistleblowers.org/html/fbiLab4-8-03.htm (last visited Sept. 21, 2004); and, John Solomon, "Panel Finds Flaws in FBI Bullet Analysis," *Charlotte Observer*, Nov. 21, 2003, at http://www.bradenton.com/mld/charlotte/news/breaking_news/7315660.htm?template=conten tModules/printstory.jsp (last visited September 21, 2004).

[132] Evans, *supra* note 131.

[133] Martin, *supra* note 131.

[134] Solomon, *supra* note 131. The 214 page report is: Committee on Scientific Assessment of Bullet Lead Elemental Composition Comparison, National Research Council. *Forensic Analysis: Weighing Bullet Lead Evidence* (Washington, DC: The National Academies Press 2004). The study is available at, http://www.nap.edu/catalog/10924.html (last visited Sept. 19, 2004).

[135] Evans, *supra* note 131.

[136] *Id.*

[137] *Id.*

[138] Murray Evans, "FBI Agent Sentenced in False Swearing Case," *The Enquirer* (Cincinnati, OH), June 18, 2003, available at, http://www.enquirer.com/editions/2003/06/18/loc_kyfbiagent18.html (last visited September 30, 2004). The Fayette County prosecutor said after her sentencing that he he had recommended, "She's already lost her job and paid severely, through the loss of her job and her reputation," Smith said. "In my mind, that was sufficient. I didn't see that the taxpayers of Fayette County needed to keep her up for a while." *Id.*

[139] *Id.*

[140] Martin, *supra* note 131. (Jacqueline Blake was unnamed in this news report as the technician involved in the 103 cases of suspect DNA analysis).

[141] *Id.*

[142] Associated Press, "FBI Reels From Wrongdoing by Workers." *St. Petersburg Times*, April 16, 2003, World/Nation section, at http://www.sptimes.com/2003/04/16/Worldandnation/FBI_reels_from_wrongd.shtml (last visited September 14, 2004).

[143] John Solomon, "Justice Broadens FBI DNA Probe," *San Francisco Examiner*, April 28, 2003.

[144] Martin, *supra* note 131.

[145] Curt Anderson, "Ex-FBI Lab Worker Guilty, Falsified DNA," *The Guardian*, May 19, 2004, at, http://www.guardian.co.uk/worldlatest/story/0,1280,-4106036,00.html (last visited Sept. 21, 2004).

[146] Solomon, *supra* note 131. The report was issued in February 2004. See, Committee on Scientific Assessment of Bullet Lead Elemental Composition Comparison, *supra* note 134.

[147] *Id.*

[148] *Id*

[149] *Id*

[150] *Id*

[151] Committee on Scientific Assessment, *supra* note 134.

[152] Solomon, *supra* note 131.

[153] *Id.*

[154] *Id.*

[155] *Id.*

[156] *Id.*

[157] In the late 1990s the FBI's database of lead test results had more than 13,000 samples. See, Solomon, *supra* note 131. Under 28 U.S.C. §§2254 and 2255, state and federal prisoners respectively, must present new evidence in order to file a successive habeas petition challenging their sentence and/or underlying conviction.

[158] Associated Press (FBI Reels), *supra* note 142.

[159] Kelly, *supra* note 25, at 2-3. (The investigation only included three of seven units in the lab's Scientific Analysis Section, and none of the lab's other 20 units.)

[160] Although there could be varying case loads by different sections, a direct extrapolation is, 3,000 x (27/3) = 27,000.

[161] The FBI crime lab's deficiencies don't mean it doesn't get "it right" in any number of cases. What it does mean is that while in some cases, such as bullet "data chaining" the FBI's testing can be categorically rejected as unsound, in other cases its evidentiary value can only be confirmed, i.e., determined to be valid by a cross-analysis of the evidence by a qualified independent forensic laboratory.

[162] Baden, *supra* note 46, at 229.

[163] *Id.* at 229.

[164] *Id.* at 229.

[165] *Id.* at 229-230.

[166] Fred Zain provided expert prosecution testimony for criminal cases at least ten states. See, *Id.* at 229-230.

[167] *Id.* at 229-230. ("Between 1977 and 1993, Fred Zain tested and testified about blood and semen evidence in hundreds of murder and rape trials."). *Id.*

[168] Kelly, *supra* note 25, at 13.

[169] Edward Connors, Thomas Lundregan, Neal Miller, Tom McEwen, "Convicted by Juries, Exonerated by Science: Case Studies in the Use of DNA Evidence to Establish Innocence After Trial," *National Institute of Justice*, 1996, at 74-76.

[170] Connors, *supra* note 169, at 74-76; see also, *In the Matter of an Investigation of the West Virginia State Police Crime Laboratory, Serology Division*, 438 S.E. 2d 501, 503; 190 W.Va. 321 (November 10, 1993), note 162, at 503 n.2 (noting that the Supreme Court, "in an order dated March 29, 1990, ... authorized the performance of a DNA test on Mr. Woodall."). *Id.*

[171] Connors, *supra* note 169, at 74-76.

[172] *What Jennifer Saw*, PBS Frontline, http://www.pbs.org/wgbh/pages/frontline/shows/dna/cotton/compensate.html (last visited Sept. 18, 2004).

[173] The 11 areas of Fred Zain's misconduct are listed in, *In the Matter of an Investigation of*

the *West Virginia State Police Crime Laboratory, Serology Division*, 438 S.E. 2d at 503. Several of the conclusions can perhaps more clearly be stated that Zain "used inaccurate math to statistically link evidence to a defendant and a certain percentage of the population." See, Rachelle Bott, "Fred Zain: the retrial," *Sunday Gazette-Mail*, Charleston, West Virginia, September 2, 2001, 1A.

[174] *In the Matter of an Investigation of the West Virginia State Police Crime Laboratory, Serology Division*, 438 S.E. 2d at 511.

[175] Janet Elliot, "Lawyers Seek Access To Bexar County Lab: New DA Reviews Zain's Work," *Texas Law*, August 2, 1999.

[176] Connors, *supra* note 169, at 34.

[177] *Id.* at 35.

[178] Kelly, *supra* note 25, at 13.

[179] Elliot, *supra* note 175.

[180] Kelly, *supra* note 25, at 13.

[181] See e.g., "Death of Lying Chemist Fred Zain," Talk Left, December 8, 2002, http://talkleft.com/new_archives/001077.html (last visited Sept. 19, 2004). (Reprinted from, "Obituaries / Fred Zain, 52, Discredited W.Va. Police Chemist," *Newsday*, Dec. 4, 2002).

[182] Zain was prosecuted on four counts of obtaining his salary and witness fees under false pretenses by lying on the witness stand and faking test results. Zain's retrial was postponed due to his colon cancer from which he died at the age of 52. See, *Id.* See also, Connors, *supra* note 169, at 18, fn. 275. See also, Robert Tanner, "Standards, Autonomy Sought for Nation's Crime Labs," *The Seattle Times*, July 7, 2003, Nation section, http://archives.seattletimes.nwsource.com/cgi-bin/texis.cgi/web/vortex/display?slug=labs07&date=20030707 (last visited Sept. 21, 2004).

[183] For an account of the difficulties encountered in Texas and West Virginia to holding Zain criminally liable for what he did, see, Bott, *supra* note 173.

[184] For an account of the difficulties encountered in Texas and West Virginia to holding Zain criminally liable for what he did, see, *Id.*

[185] "Lost, Mishandled Evidence May Mean New Appeals," *Florida Prison Legal Perspectives*, Vol. 6, Issue 6, at: http://www.vaccinationnews.com/DailyNews/2003/May/12/LostMishandled12.htm (last visited on September 14, 2004).

[186] *Id.*

[187] See e.g., Francine Yurko, "Triumph Over Tragedy," *Justice Denied*, Issue 23, Winter 2004, 10-11. See also, Hans Sherrer, "Baby Alan's Mysterious Death: Alan Yurko is innocent, so who is responsible for his son's death?," *Forejustice.org*, Oct. 15, 2003, http://forejustice.org/wc/baby_alan.htm. (last visited September 14, 2004).

[188] *Id.*

[189] See, Anthony Colarossi and Pamela J. Johnson, "Dad freed from life sentence in son's death," *Orlando Sentinel*, August 28, 2004.

[190] *Id.*

[191] Yurko, *supra* note 187, at 10-11.

[192] Amy C. Rippel and Anthony Colarossi, "Dead Baby's Dad Closer to Retrial," *Orlando Sentinel*, March 27, 2004.

[193] Amy C. Rippel and Henry Pierson Curtis, "Criticism, Blunders Mar Medical Examiner's Exit," *Orlando Sentinel*, August 31, 2003.

[194] *Id.*

[195] *Id.*

[196] *Id.*

[197] *Id.*

[198] Associated Press, "Inmate Freed; Cleared in sex assault case after 15 years," *St. Augustine Record*, May 7, 2001, at http://www.staugustine.com/stories/050801/nat_0508010029.shtml (last visited September 14, 2004).

[199] *Id.* See also, Pamela Gould, "2 Labs, 2 Cases, 2 Goofs," *The Free Lance-star* [Fredericksburg], May 29, 2001, at http://www.fredericksburg.com/News/FLS/2001/052001/05292001/288205 (last visited September 14, 2004). This article about Jeffrey Pierce's exoneration relates that in 1997, a Virginia man was in jail awaiting trial for the murder of Sofia Silva based on the Virginia state crime lab's analysis that fibers found in a search of his vehicle matched fibers found at the scene of her murder. *Id.* While jailed, two other women were murdered in a similar fashion to Ms. Silva, and fibers were found at their murder scene matching those found at the scene of Ms. Silva's murder. *Id.* The suspect in her murder was excluded upon reexamination of the fibers found in his vehicle. *Id.*

[200] Associated Press (Inmate Freed), *supra* at note 198.

[201] *Id.*

[202] Jim Yardley, "Inquiry Focuses on Scientist Used by Prosecutor," *The New York Times*, May 2, 2001.

[203] *Id.*

[204] *Id.*

[205] *Id.*

[206] *Id.*

[207] *Id.*

[208] *Id.*

[209] *Id.*

[210] *Id.*

[211] *Id.*

[212] *Id.*

[213] *Id.*

[214] Randy Ellis, "Forensic Chemist's Work Criticized," *The Oklahoman*, September 15, 1999, at http://www.law-forensic.com/cfr_gilchrist_8.htm (last visited September 14, 2004).

[215] *See* Hans Sherrer, "The Innocents: the Prosecution, Conviction, and Imprisonment of the Innocent, Introduction (Part One)," *Justice Denied*, Vol. 1, Issue 2, p. 32. That estimate is supported by a detailed analysis that over 14 percent of all convictions in state and federal courts are of innocent people. *See* Hans Sherrer, *How Many Innocents Are There?*, 43 (Feb. 8, 2003) (Unpublished manuscript, on file with the author.) This estimate was made prior to the publishing of a study that included all 4,578 capital appeals finalized in the U.S. between 1973 and 1995. James Liebman, Jeffrey Fagan & Valerie West, "A Broken System: Error Rates in Capital Cases and What Can be Done About Them," *The Justice Project*, June 12, 2000. Report available in its entirety from The Justice Project: http://justice.policy.net/jpreport/ (last visited September 14, 2004). (Overseen by Columbia University School of Law Professor James Liebman, the study found that, "7% of capital cases nationwide are reversed because the condemned person was found to be innocent.") *Id.*

[216] Tanner, *supra* note 182.

[217] *Id.*

[218] For a concise summary of the prosecution's evidentiary problems in the O.J. Simpson case, see, Kelly, *supra* note 25, at 231-270 (Chapter 7, O. J. Simpson: Dirty Hands, Bad Blood).

[219] See e.g., *Id.* at 238. ("Yet the evidence itself was becoming more overwhelming by the day. . … A further inoculation against defense challenges was to get all three labs to use both available DNA testing methods, the lengthier, more reliable RFLP method as well as the newer, less certain PCR test.") *Id.*

[220] See e.g., Cole, *supra* note 1, at 300. ("They attacked the weakest link in the processing of the DNA evidence, the work of the LAPD's forensic technicians Dennis Fung and Andrea Mazzola, who had committed numerous procedural errors in recovering, storing, and transporting the evidence to the LAPD crime laboratory. … In his closing argument Scheck compared the hygiene of the LAPD's forensic evidence truck to a New York City restaurant in which cockroaches are visible, and he suggested that the evidence implicating Simpson had been contaminated or, worse, planted by the police themselves.") *Id.*

[221] Kelly, *supra* note 25, at 232-233. (An example of the sloppy evidence tracking is that in August 1994 "the LAPD had no standard system of record collection for evidence." *Id.* at 240. It is interesting that the FBI's lab was not involved in testing the DNA evidence in the O.J. case for several reasons. First, Judge Ito ruled that defense observers could be present during DNA testing which the FBI would not allow. *Id.* at 238. Second, the FBI used a method of calculating random match probabilities that had not been accepted by California state courts. *Id.* at 238. Third, the FBI's method of testing DNA was considered to be too subjective for acceptance by California state courts. *Id.* at 238-239. In other words, if the FBI crime lab had been involved, the deficiencies in the handling and testing of the DNA evidence in O.J.'s case can be expected to have been significantly more egregious than they were revealed to have been.)

[222] *Id.* at 242 (Photographs taken of a gate on June 13, 1994, the day after the murder of Nicole Brown Simpson and Ronald Goldman, did not show any blood stains. However on July 3, 1994 blood samples allegedly from O.J. Simpson were collected from the gate by the LAPD. *Id.* "And since the [blood] stains on the gate, in the Bronco, and on the sock had been missed until three weeks after the murders, the defense could argue that they had not been there during the initial searches." *Id.* at 240.)

[223] *Id.* 232-233. An example of the sloppy evidence tracking is that in August 1994 "the LAPD had no standard system of record collection for evidence." Id. at 240. It is interesting that the FBI's lab was not involved in testing the DNA evidence in the O.J. case for several reasons. First, Judge Ito ruled that defense observers could be present during DNA testing which the FBI would not allow. *Id.* at 238. Second, the FBI used a method of calculating random match probabilities that had not been accepted by California state courts. *Id.* at 238. Third, the FBI's method of testing DNA was considered to be too subjective for acceptance by California state courts. *Id.* at 238-239. In other words, if the FBI crime lab had been involved, the deficiencies in the handling and testing of the DNA evidence in O.J.'s case can be expected to have been significantly more egregious than they were revealed to have been.

[224] See e.g., *Id.* 233-4, 241.

[225] *Id.*

[226] O.J.'s attorney Barry Scheck made that comparison. See, Cole, *supra* note 1, at 300.

[227] *Id.* 300-301.

[228] Kelly, *supra* note 25, at 233.

[229] Baden, *supra* note 46 at 229-230.

[230] Michael Daecher, "Forensic Fraud: No Motive, No Witness, and 99 Years in Jail: The Curious Conviction of Sonia Cacy," *Texas Observer*, August 28, 1998, at: http://www.law-forensic.com/cfr_arson_cacy.htm (last visited September 14, 2004). See also, Janet Elliot, "Lawyers seek access to Bexar County Lab; New DA reviews Zain's work," *Texas Lawyer*, Aug. 2, 1999, at 1.

[231] *Id.*

[232] Daecher, *supra* note 230.

[233] *Id.*

[234] *Id.*

[235] *Id.* ("According to the autopsy report, Richardson's lungs were not filled with smoke or

soot, nor was there excessive carbon monoxide in his blood (his carbon monoxide levels were about 11 percent, consistent with heavy smoking, while two dogs who died in the fire had carbon monoxide levels of about 40 percent and 60 percent. ... Nowhere in the autopsy report is there any mention of soot in the trachea or larynx, and it appeared that Richardson was dead before the fire started."). *Id.* See also, Elliot, *supra* note 230.

[236] Daecher, *supra* note 230. (The Wall Street Journal and ABC News were among the national news media that reported on irregularities in Sonia Cacy's case.)

[237] Connors, *supra* note 169, at 34.

[238] *Id.* at 34.

[239] *Id.* at 34.

[240] *Id.* at 35.

[241] *Id.* at 35.

[242] *Id.* at 34.

[243] Tanner, *supra* note 182.

[244] *Id.*

[245] *Id.*

[246] See, Jaci Webb and Greg Tuttle, "Free After 15 years," *Billings Gazette*, Oct 2, 2002.

[247] *Id.*

[248] *Id.*

[249] *Id.*

[250] *Id.*

[251] *Id.*

[252] *Id.*

[253] *Id.*

[254] Lise Olsen, "Crime Lab Worker Failed To Qualify To Test Hair Samples," *Seattle Post-Intelligencer*, January 2, 2002, at, http://seattlepi.nwsource.com/local/102492_lab02.shtml?searchpagefrom=1&searchdiff=370. (last visited September 15, 2004).

[255] *Id.*

[256] *Id.*

[257] Olsen, *supra* note 254.

[258] *Id.*

[259] Ruth Teichroeb, "Shadow of Doubt," *Seattle Post-Intelligencer*, March 13, 2004, A1, A7.

[260] Olsen, *supra* note 254.

[261] Ruth Teichroeb, "State Patrol fires crime lab scientist: His testimony in Montana cited; internal audit is downplayed," *Seattle Post-Intelligencer*, March 24, 2004, B1, B5, at B5.

[262] Tanner, *supra* note 182.

[263] *Id.*

[264] *Id.*

[265] *Id.*

[266] *Id.*

[267] *Id.*

[268] Olsen, *supra* note 254.

[269] Teichroeb, *supra* note 259, at A1, A6-7.

[270] *Id.* (The audit report was obtained through Washington State Public Records Act.)

[271] *Id. at* A6.

[272] *Id.* at A1, A6-7.

[273] *Id. at* A6.

[274] *Id. at* A6.

[275] *Id. at* A1, A6-7.

[276] *Id.* at A1, A6-7.

[277] Teichroeb, *supra* note 261, at B5.

[278] For explanations of the results of crime lab personnel proficiency tests from 1974 to 2001, see *supra*, Chapter 2: Shoddy Work is the Norm for Crime Labs.

[279] Teichroeb, *supra* note 259, at A7. (Those three men, all wrongly convicted of rape, were Chester Bauer, Jimmy Ray Bromgard and Paul Kordonowy. Those men were respectively wrongly imprisoned for 14, 15 and 13 years.) *Id.*

[280] John K. Wiley, "State Patrol fires embattled crime-lab scientist," *The Seattle Times*, March 24, 2004, page B3. http://seattletimes.nwsource.com/html/localnews/2001886614_crimelab24m.html (last visited September 15, 2004). ("Forensic scientists with expertise in fiber and hair examinations later concluded Melnikoff's testimony on the number of hair examinations he had conducted and statistical comparisons contained "egregious misstatements.""). *Id.* Paul Kordonowy was prosecuted in Montana, and Melnikoff's testimony in the case was based on work he allegedly performed while employed at the MSPCL. *Id.*

[281] Teichroeb, *supra* note 259, at A1, A6-7. (Details Melnikoff's suspect conduct while working at the WSP crime lab.). Melnikoff's substandard lab work is only the tip of the iceberg of the low quality of work that has been performed by WSP's crime lab technicians for decades. For an in-depth investigative report of proficiency problems at the lab dating from the mid-1970s to 2004, see, Ruth Teichroeb, "Oversight of crime-lab staff has often been lax," *Seattle Post-Intelligencer*, July 23, 2004, at, http://seattlepi.nwsource.com/local/183203_crimelab23.html?searchpagefrom=1&searchdiff=69 (last visited September 30, 2004).

[282] Tanner, *supra* note 182.

[283] *Id.*

[284] *Id.*

[285] Kelly, *supra* note 25, at 13. In sort of a bizarre follow up to the escapades of Dr. Erdmann, in Oct. 1, 2003 it was reported that Dr. Jerry Spencer, the Chief Medical Examiner for Lubbock County, one of the countries that Erdmann was ME for, was under investigation for robbing female corpses of their breast implants. Kerry Drennan, "Tech Reviews Allegations Against Medical Examiner," *Amarillo Globe-News*, Oct. 1, 2003, at: http://www.amarillonet.com/stories/100103/tex_allegations.shtml (last viewed Sept. 17, 2004).

[286] Kelly, *supra* note 25, at 13.

[287] *Id.* at 13.

[288] Underwood, *supra* note 4, at 174-175.

[289] *Id.* at 174-175.

[290] *Id.* at 174-175.

[291] *Id.* at 174-175.

[292] Graham Underwood, "Freedom nears for Erdmann as judge clears way for parole," *Lubbock Avalanche-Journal*, February 26, 1997, http://www.lubbockonline.com/news/022797/freedom.htm (last viewed September 17, 2004).

[293] Kelly, *supra* note 25, at 13.

[294] Underwood, *supra* note 4, at 175.

[295] *Id.* at 174 and note 121.

[296] *Id.* at 175; See also, Kelly, *supra* note 25, at 13.

[297] W. B. Allen, "The Laws of Science and the Rule of Law." A Presentation in the Forum "Science, Ethics and the Law" for the *Eighth Conference on Statistics, Science and Public Policy*, April 23-26, 2003, at: http://williambarclayallen.com/presentations/Laws%20of%20science%20and%20the%20rule%20of%20law.htm (last visited September 17, 2004). Louise Robbins died in 1987.

[298] Ray Gibson, "Courted Expert Steps on Toes with Footprints," *Chicago Tribune*, April 6, 1986, p. 1.

[299] Mark Hansen, "Believe It Or Not," 79 *A.B.A.J.* 64 (June 1993), at: http://www.science.sjsu.edu/bio101/believe_it_or_not.htm (last viewed September 17, 2004) (Hansen also wrote, "Many of her colleagues in the field of anthropology echoed these remarks, noting that at an archeological that at an archeological dig in Tanzania she had misidentified one set of human prints as belonging to an antelope and had made the totally unfounded claim that another print was made by a prehistoric woman who was 5½ months pregnant. Her conclusions were dismissed as all "in her mind.") *Id.*

[300] *Id.*

[301] Kelly, *supra* note 25, at 13.

[302] *Id.* at 13.

[303] Mark Hansen, "Out of the Blue," 82 *A.B.A.J.* 50 (February 1996), at 50, 51, quoted in, Craig M. Cooley, "Forgettable Science, Forensic Science, and Capital Punishment: Reforming the Forensic Science Community to Avert the Ultimate Injustice," 15 *Stan. L. & Pol'y Rev* 2, (2003), online at: http://www.law-forensic.com/final_III.htm, p 12, fn 190. (last viewed September 17, 2004).

[304] Kelly, *supra* note 25, at 13.

[305] *Id.* at 13.

[306] *Id.* at 13.

[307] *Id.* at 14.

[308] Hansen, supra note 303 at 50. (Quoted in Kelly, *supra* note 25, at 14.)

[309] Tamara Audi, "Bones of Contention: Cadaver-sniffing canine's finds are under suspicion; Handler of world-famous dog is charged with planting evidence," *Detroit Free Press*, July 14, 2003, http://www.freep.com/news/mich/dog14_20030714.htm (last visited Sept. 21, 2004).

[310] *Id.*

[311] *Id.*

[312] Associated Press, "Cadaver Hunter Is Indicted: Woman is accused of planting evidence during victim searches," *The Detroit News*, Metro/State section, August 21, 2003, at, http://www.detnews.com/2003/metro/0308/21/d11d-249816.htm (last visited Sept. 21, 2004).

[313] *Id.*

[314] Anderson's method of systematically creating the illusion of Eagle's unusual talent is reminiscent of how prosecutor's use tactics of illusion to make an innocent person appear guilty of a criminal offense. See e.g., Hans Sherrer, "The Illusion of Tax Evasion," in *The Ethics of Tax Evasion*, Robert W. McGee, ed. (The Dumont Institute for Public Policy Research , 1998), 102-122.

[315] Richard Luscombe, "Canine Sleuth Finds His Record Blighted," *The Observer* (London, UK), Sept. 21, 2003, http://observer.guardian.co.uk/international/story/0,6903,1046383,00.html (last visited Sept. 21, 2004).

[316] *Id.*

[317] *Id.*

[318] *Id.*

[319] Audi, *supra* note 309.

[320] *Id.*

[321] Christopher M. Singer, "Cadaver-dog Case May Boost Appeals," *The Detroit News*, August 22, 2003, http://www.detnews.com/2003/wayne/0308/22/d03-251184.htm (last visited September 21, 2004).

[322] *Id.*

[323] Audi, *supra* note 309.

[324] Associated Press (Cadaver Hunter), *supra* note 312.

[325] Luscombe, *supra* note 315.

[326] Jack Kresnak, "Bones-finder Says She Cheated," *Detroit Free Press*, March 12, 2004, http://www.freep.com/news/mich/dog12_20040312.htm (last visited September 21, 2004).

[327] David Shepardson, "Dog-handler Pleads Guilty," *The Detroit News*, March 12, 2004, http://www.detnews.com/2004/metro/0403/12/d01-89790.htm (last visited Sept. 21, 2004).

[328] Associated Press, "Michigan Cadaver Dog Handler Pleads Guilty to Federal Charges," *Lincoln Journal-Star*, March 11, 2004. At, http://journalstar.com/articles/2004/03/12/nebraska/10046626.txt (last visited Sept. 21, 2004).

[329] David Ashenfelter, "Renowned Dog Handler Sentenced to 21 Months," *Detroit Free Press*, September 29, 2004, at http://www.freep.com/news/metro/anderson29e_20040929.htm (last visited September 30, 2004).

[330] Christopher M. Singer, "Retrial Order to be appealed," *The Detroit News*, August 15, 2004, at, http://www.detnews.com/2004/wayne/0408/16/b04-242390.htm (last visited September 30, 2004).

[331] Judge Orders Retrial For Convicted Murderer, *Newstalk WJR 760am* [Detroit, MI], July 2004. At http://www.760wjr.com/Article.asp?PT=Local+News&id=33373 (last visited September 30, 2004).

[332] The frauds perpetrated by Sandra Anderson were discovered by people in the field who observed her planting bones. Laboratory corroboration of the insubstantial nature of Ms. Anderson's alleged finds only occurred subsequent to those visual observations. *Id.*

[333] See e.g., Oskar Pfungst, *Clever Hans (The Horse of Mr. Von Osten): A Contribution To Experimental Animal And Human Psychology* (Henry Holt and Company 1911). This book is the single best source of information about Clever Hans, since it was written by the researcher who discovered the secret of his ability to correctly solve math problems and many other "cognitive" feats, by tapping the answer with his right forefoot.

[334] Scott Glover and Matt Lait, "Credibility of Hollywood private eye shattered," *The Seattle Times*, February 2, 2004, A7.

[335] *Id.*

[336] *Id.*

[337] *Id.*

[338] *Id.*

[339] *Id.*

[340] *Id.*

[341] *Id.*

[342] *Id.*

[343] *Id.*

[344] *Id.*

[345] *Id.*

[346] *Id.*

[347] *Id.*

[348] Maurice Possley, "Ex-Inmate Exonerated of Rapes Tries To Get His Life In Order," *Chicago Tribune*, June 29, 2000, A4, http://www.law-forensic.com/cfr_fish_13.htm (last visited September 21, 2004).

[349] Steve Mills and Maurice Possley, "Report alleges crime lab fraud: Scientist is accused of providing false testimony," *Chicago Tribune*, January 14, 2001, http://www.chicagotribune.com/news/specials/chi-010114roscetti,0,2442127.story?coll=chi-newsspecials-utl (last visited September 21, 2004).

[350] Barry Scheck, Peter Neufeld and Jim Dwyer, *Actual Innocence* (Penguin Putnam 2001 rev. ed.), 160. (Noting those assaults occurred in the same neighborhood of Chicago as the beauty

shop rapes and followed the same MO, except the rapist targeted bars instead of beauty parlors.).

[351] Mills, *supra* note 349.

[352] Possley, *supra* note 348.

[353] Frank Main and Carlos Sandovi, "2 Charged in Roscetti Case," *Chicago Sun-Times*, February 8, 2002, http://www.kathleentzellner.com/news35.html (last visited September 19, 2004).

[354] Maurice Possley, "Crime lab defends itself, analyst: City unit cited in allegations," *Chicago Tribune*, July 20, 2001, http://www.chicagotribune.com/news/nationworld/chi-010720roscetti,1,3979533.story (last visited September 19, 2004).

[355] Steve Mills, Maurice Possley and Kim Barker, "3 Roscetti inmates walk free: After 15 years, new world greets them as judge tosses convictions," *Chicago Tribune*, December 6, 2001, at 1. (In exchange for a confession that turned out to be false, and an agreement to testify for the prosecution against Larry Ollins, Marcellius Bradford pled guilty to a lesser charge and was released after serving 6-1/2 years of a 12 year sentence.), at, http://www.chicagotribune.com/news/nationworld/chi-011206roscetti,1,3389709.story (last visited September 15, 2004).

[356] Margaret Cronin Fisk, "Lawyers Free Chicago Trio After Retesting of Lab Sample: Technician Had Testified in Earlier Bad Conviction," *Nat'l L.J.*, December 17, 2001.

[357] Mills, *supra* note 349.

[358] Fisk, *supra* note 356.

[359] *Id.*

[360] *Id.*

[361] *Id..*

[362] Mills, *supra* note 355, at A1. (Larry Ollins, Calvin Ollins and Omar Saunders were all released in December 2001 after serving almost 15 years of their life sentences, while Marcellus Bradford, served 6-1/2 years after making a deal to falsely testify against Larry Ollins.).

[363] Maurice Possley and Steve Mills, "Governor pardons Roscetti 4: Action paves way for ex-inmates to get compensation from state fund," *Chicago Tribune*, October 17, 2002. http://www.chicagotribune.com/news/nationworld/chi-021017roscetti,1,4045070.story (last visited September 21, 2004).

[364] Fisk, *supra* note 356.

[365] Mills, *supra* note 349.

[366] *Id.*

[367] *Id.*

[368] *Id.*

[369] Maurice Possley and Ken Armstrong, "Lab Tech in Botched Case Promoted: Testimony Helped Wrongfully Convict Man of Rape," *Chicago Tribune*, March 19, 1999, Metro Chicago Section, p. 1.

[370] Possley, *supra* note 354.

[371] Mills, *supra* note 349.

[372] *Id.* (Recounting an interview with Dr. Edward Blake.).

[373] For a more complete account of Josiah Sutton's case, see *infra* Chapter 7: DNA – Probability Estimates Elevated by Smoke and Mirrors to Certainty.

[374] See e.g., Press Release, "UCI Professor Finds DNA Laboratory Errors That Sent Wrong Man to Prison," *Scientific Testimony: An Online Journal*, March 11, 2003, at: http://www.scientific.org/archive/Sutton%20Press%20Release.htm (last visited Sept. 20, 2004).

[375] Lise Olsen and Roma Khanna, "DNA lab analysts unqualified: Review Finds Education,

Training Lacking," *Houston Chronicle*, September 7, 2003, at: http://www.chron.com/cs/CDA/printstory.hts/special/crimelab/2085350 (last visited September 21, 2004). (The lab was anticipated to resume operations sometime in 2004.).
[376] Press Release ("UCI Professor"), *supra* note 374.
[377] *Id.*
[378] *Id.*
[379] *Id.* ("An audit report, released in January [2003], confirmed Thompson's claim that the laboratory was employing dangerous and inappropriate procedures...") *Id.*
[380] Olsen, *supra* note 375.
[381] Thirteen of the 49 cases revealed significant problems with the HPD Crime Lab's work. See, Roma Khanna, "Credentials embellished: Transcript: Ex-lab chief told jury he had a Ph.D.," *Houston Chronicle*, September 10, 2003, at, http://www.chron.com/cs/CDA/printstory.hts/special/crimelab/2090966, (last visited September 21, 2004). A private lab was also unable to replicate the crime lab's test results in multiple cases. See, Roma Khanna and Steve McVicker, "Private DNA lab can't replicate HPD results," *Houston Chronicle*, September 5, 2003, at, http://www.chron.com/cs/CDA/printstory.hts/special/crimelab/2035462 (last visited September 21, 2004).
[382] Steve McVicker and Roma Khanna, "29 cases missing evidence: 8 added to tally at police DNA lab," *Houston Chronicle,* November 12, 2003, p. 1, at, http://www.chron.com/cs/CDA/ssistory.mpl/special/crimelab/2219005 (last visited September 15, 2004). (The defendant's deprived of DNA tests had been convicted of various crimes, including one of murder and several of sexual assault.)
[383] Roma Khanna and Steve McVicker, "Report outlines steps to validate crime lab," *Houston Chronicle*, August 7, 2003, at, http://www.chron.com/cs/CDA/printstory.hts/special/crimelab/2035462 (visited last on September 21, 2004).
[384] This is demonstrated by the fact that proficiency testing, even if it were associated with accreditation (which it typically is not), does not necessarily contribute to ensuring the *accuracy* of test results. See e.g., Thompson, *supra* note 10, at 48. ("...this work [proficiency testing] is designed more to test the uniformity of DNA test results among laboratories using the same protocol than to determine the rate of errors.") *Id.*
[385] Adam Liptak, "Prosecutions Are a Focus in Houston DNA Scandal," *New York Times*, June 9, 2003, available at, http://www.truthinjustice.org/houston-da.htm (last visited September 21, 2004).
[386] Roma Khanna, "New Evidence Furor Hits HPD," *Houston Chronicle*, August 27, 2004, A1.
[387] *Id.*
[388] See e.g., Underwood, *supra* note 4, at 152.
[389] Thompson, *supra* note 10, at 47. ("A false positive might occur due to error in the collection or handling of samples, misinterpretation of test results, or incorrect reporting of test results.")
[390] Teichroeb, *supra* note 261, at B5, and quote in accompanying text. (WSP crime lab technician Arnold Melnikoff's 30% error rate in tests of physical evidence "wouldn't pass a first-year college chemistry class."). *Id.* Although such a level of proficiency might be disturbing if exhibited by high school students seeking the approval of their teacher or a higher grade than they deserve, it is much more than that when engaged in by people who hold the fate of men and women in their hands – everyone of whom is presumed under the law to be innocent. *In re Winship*, 397 U.S. 358, 363-364 (1970). The degree of the irregularities in the FBI's crime lab over a period of almost three decades is recounted at length in Kelly,

supra note 25.

[391] See e.g., *In re Winship*, 397 U.S. at 363-364.

[392] The only reason their has been a lack of consistent exposure by the national media of suspect police evidence collection and crime lab operation irregularities such as occurred in the O.J. Simpson case, is DNA evidence is typically used against low-profile people of limitied means who must rely on the limited investigative resources available to their public defender.

[393] See e.g., Underwood, *supra* note 4, at 175-176.

[394] *Id.* at 152. ("...forensic scientists may be over-zealous or manipulated by their prosecutorial "customers...")

[395] Audi, *supra* note 309.; and, Associated Press, *supra* note 312.

[396] Scott Glover and Matt Lait, "Even prosecutors sought pellicano for his expertise," *Los Angeles Times*, February 1, 2004, at A1.

[397] Kelly, *supra* note 25, at 13.

[398] Underwood, *supra* note 4, at 152, fn 11, See also, Richard Underwood, ""X-Spurt" Witnesses," 19 *Am. J. Trial Advoc.* 343 (1995).

[399] *In re Winship*, 397 U.S. 358, 363-364 (1970) (citations omitted).

[400] For a review of the multitude of tactics used to frame an innocent person, see: Sherrer, *supra* note 34; see also, *Prosecutorial Lawlessness, supra* note 34.

[401] For a review of the multitude of tactics used to frame an innocent person, see: *Id.*.

[402] James McCloskey, "Convicting the Innocent," *Voice for the Defense*, December, 1989, at 20-25. ("Another problem that we continually observe within the realm of forensic evidence is the phenomenon of lost and untested physical evidence. Often, especially in cases up to the early 1980's, the specimens that have the potential to exclude the defendant have not been tested and eventually get misplaced. At best this is gross negligence on the part of both the police technician and the defense attorney in not ensuring that the tests be done.") *Id.*

[403] *Id.* at 20-25. ("Another problem that we continually observe within the realm of forensic evidence is the phenomenon of lost and untested physical evidence. Often, especially in cases up to the early 1980's, the specimens that have the potential to exclude the defendant have not been tested and eventually get misplaced. At best this is gross negligence on the part of both the police technician and the defense attorney in not ensuring that the tests be done.") *Id.*

[404] *Id.* at 20-25. See also e.g., Kelly, *supra* note 25, at 14. (relating how FBI agent Tom Curran, who worked as an examiner in the FBI lab, "issued reports of blood analyses when "no laboratory tests were done"' had relied on presumptive tests to draw up confirmatory results; and had *written up inadequate and deceptive lab reports, ignoring or distorting test results.*") (emphasis added) *Id.*

[405] *Arizona v. Youngblood*, 488 U.S. 51 (1988).

[406] *Id.* at 58.

[407] *Id.* at 52, 54.

[408] *Id.* See, *State v. Youngblood*, 734 P.2d 592, 153 Ariz. 50 (1986).

[409] *Id.* at 61, 73. (Justice Blackmun dissenting)

[410] Scheck, supra note 350, at 334-36.

[411] *Id.* at 334-36.

[412] *Id.* at 334-36.

[413] *Youngblood*, 488 U.S. 51.

[414] See e.g., Hans Sherrer, "The Complicity of Judges in the Generation of Wrongful Convictions," 30 *N. Ky. L. Rev.* 539, 569-570 (2003) (Discussing how the Supreme Court's *Youngblood* ruling excusing prosecution related errors reflects a liberal application of the 'harmless error' rule.).

[415] *Illinois v. Gregory Fisher*, 124 S.Ct. 1200 (2004).

[416] *Id.* at 1202. ("We have never held or suggested that the existence of a pending discovery request eliminates the necessity of showing bad faith on the part of police.") *Id.* (It is notable that both *Youngblood* and *Fisher* were reversals of state due process decisions that provided a shield of protection for an innocent person from a wrongful conviction.).

[417] *Id.* at 1202 ("At most, respondent could hope that, had the evidence been preserved, a fifth test conducted on the substance would have exonerated him.") *Id.*

[418] *Id.* at 1202.

[419] See e.g., Yant, *supra* note 20, at 66, (At a 1985 symposium on forensic chemistry, Professor Benjamin W. Grunbaum deplored the "orientation of the analyst within the criminal-justice system," whose pro-prosecution bias "makes it difficult to maintain scientific objectivity." He also reasoned a "substantial" number of wrongful convictions could result from crime lab practices and the bias of their technicians.). *Id.*

[420] The strongest support for the continued acceptance of fingerprint evidence's probative value is it has been accepted by courts for so long (in this country since 1910). Cole, *supra* note 1, at 179. *Daubert* based challenges to fingerprint evidence have thus far been unsuccessful because they have been unable to overcome the inertia of judges, who although they may not come out and say it, express in their opinions that what has been good enough to be considered as evidence in the past, should continue to be considered good enough in the present. It is hoped that articles such as this contribute to a reevaluation of that 'head in the sand' attitude.

[421] The probability of a given roll of two dice ranges from a low of .0278 for a two or twelve, to a high of .1667 for a seven. See, Wilf Hey, "Random Numbers," *PC Plus*, April 2004, at 176-178, esp. 178 (Explaining how to write a computer program that will generate a random number approximating the probability of a random roll of two dice.) As explained in *supra* notes 60-89 and accompanying text, the individual error rate of the easier than real life fingerprint examiner proficiency tests from 1983 to 2001 exceeded the 16.67% probability of randomly rolling a seven.

[422] Which means it actually aids a defendant, since a shadow is cast over the prosecution's case when what may be the most significant alleged evidence against a defendant turns out to have no probative value.

[423] While not set out in an itemized fashion as is done in the text, the essence of those three presumptions is set forth in Cole, *supra* note 1, at 175-177.

[424] Cole, *supra* note 1, at 90.

[425] This relates to *methodology* error. In 1999 FBI supervisory fingerprint specialist Meagher testified in a pretrial hearing in *U.S. v. Mitchell*, that related to the scientific *methodology* used to analyze fingerprints, the error rate was "zero." *U.S. v. Mitchell*, CR No. 96-407 (E.D. PA)). (Testimony of FBI technician Meagher, Tr. July 8, 1999, at 154-56),

[426] This relates to *practitioner* error. See e.g., the 1910 testimony of an examiner in the first criminal case in this country that relied on fingerprint evidence to support a conviction:

"Q. In comparing these fingers it is your opinion that the lines in those photographs were made by the same person?
A. I am positive. It is not my opinion." *People v Jennings* (Ill. 1910), trial transcript at 137-139. Quoted in Cole, *supra* note 1, at 179.

That attitude of the blind belief in the exactness of fingerprint testimony remains to this day. For example, during a 1999 pretrial hearing in *U.S. v. Mitchell*, a FBI lab technician testified that "*practitioner* error can be detected and corrected by another qualified examiner," prior to being testified to in court. *U.S. v Llera Plaza*, No. 98-362-10 (E.D.Pa. 01/07/2002); 2002.EPA.0000003 ¶ 114 < http://www.versuslaw.com>, quoting *U.S. v. Mitchell*, CR No. 96-407 (E.D. PA)). Test. Budowle, Tr. July 9, 1999, at 122-123, quoted in Gov't Mot. & Resp. at 42-43.

[427] In federal court under FED. R. EVI. 702, and in state courts under comparable rules.
[428] In federal court under Fed. R. Evi. 403 and/or 404, and in state courts under comparable rules.
[429] Cole, *supra* note 1, at 177-180. (Discusses the Illinois Supreme Court's 1911 case of *People of Illinois v. Jennings*, 252 Ill. 534, 96 N.E. 1077 (1911). It was the first case in this country in which a conviction dependent on fingerprint testimony tying a defendant to a crime was upheld by an appellate court. Jennings was subsequently hanged on February 16, 2012.

In the early 1900s appeals courts in countries other than the U.S. affirmed the admittance of expert fingerprint testimony based on the assumption a person's fingerprints are unique.

In the 1904 Indian case of *Emperor v. Sahdeo*, 3 Nagpur, L.Rep. 1 (1904) the court ruled, "The papillary ridges presented by the surface of the skin on the palms of the hand and soles of the feet, have been ascertained to be the most important of anthropological data." The court decided that if competent expert testimony proved two finger impressions made at different times, in different places, contain several points of agreement and no points of disagreement in their ridge characteristics, no further evidence was necessary to prove the impressions were made by the same finger. The Indian court in *Emperor* expressed approval of the principle each person's fingerprints are unique without any proof of its assertion that, "there are no two human beings in the world who exactly resemble one another in every single detail. ... "the absence of absolute repetition seems to be a universal law of nature.""

In England the Court of Criminal Appeal ruled in 1909 that expert fingerprint testimony can be received in evidence. *In Re Castleton*, 3 Crim. App. R. 74 (Cohen 1909) the Court refused to interfere with a conviction even though the fingerprint evidence was the sole basis of the defendant's identification. In 1905 Alfred Stratton was the first person convicted in England solely based on fingerprint identification. One of his fingerprints was "matched" to a bloody print found on a cash box at the scene of a double murder. His brother Albert was convicted as his accomplice, although there was no fingerprint evidence used to tie him to the crime. The judge wasn't convinced the two men were guilty, but he was bound by the jury's verdict and sentenced the two men to hang. Cole, *supra* note 1, at 172-174. It wasn't until two years after Stratton's conviction that England's Criminal Court of Appeal was established, and it ruled *In Re Castleton* on the admissibility of expert fingerprint evidence.

In Australia the Supreme Court of Victoria recognized the admissibility of expert fingerprint testimony in the 1912 case of *Parker v. Rex*, (Australia) 14 Commw. L.R. 681; 3 B.R.C. 68 (1912) ("...A fingerprint is therefore in reality an unforgeable signature.").

For additional information about early fingerprint cases see: Andre A. Moenssens, "Admissibility of Fingerprint Evidence and Constitutional Objections to Fingerprinting Raised in Criminal and Civil Cases," *Chicago-Kent Law Review*, Vol. 40, Issue 2, 85(39), Oct. 1963, 86-94.
[430] Sharath Pankanti, Salil Prabhakar and Anil K. Jain, "On The Individuality of Fingerprints," *IEEE Transactions on Pattern Analysis and Machine Intelligence*, Volume 24, Issue 8 (August 2002) at 3, article available at, http://biometrics.cse.msu.edu/prabhakar_indiv_pami.pdf (last visited September 28, 2004). Also, the uniqueness of fingerprints is called into doubt by unanswered questions such as that raised by testimony in a 1912 German case that hereditary relationships can make aspects of two or more people's fingerprints indistinguishable from each other. Cole, *supra* note 1, at 102.
[431] See e.g., Henry C. Lee and R.E. Gaensslen, editors, *Advances in Fingerprint Technology*, Second Edition (CRC Press 2001). ("From a statistical viewpoint, the *scientific foundation* for fingerprint individuality is incredibly weak.") *Id.* at 383.
[432] Pankanti, *supra* note 430, at 3.
[433] Lee, *supra* note 431, at 383. ("From a statistical viewpoint, the *scientific foundation* for

fingerprint individuality is incredibly weak." (emphasis added)). *Id.*
[434] *Id.* at 383. (emphasis added).
[435] See e.g., *Id.* at 383. See also, Pankanti, *supra* note 430, at 1, 3.
[436] See e.g., Haber, *supra* note 60, at 13. ("Friction ridge identifications are absolute conclusions. Probable, possible, or likely identifications are outside the acceptable limits of the science of friction ridge identifications.") *Id.*
[437] The first modern use of fingerprints (or more precisely a hand print) was in British ruled India in 1858. See, COLE, *supra* note 1, at 65.
[438] See e.g., *Id.* at 64-73.
[439] See e.g., *Id.* at 79. ("These [Galton points] were points along a papillary ridge where the ridge ended abruptly, split in two, or split and then rejoined.") *Id.*
[440] *Id.* at 81.
[441] See e.g., *Id.* at 79. ("These [Galton points] were points along a papillary ridge where the ridge ended abruptly, split in two, or split and then rejoined.") *Id.*
[442] See, *Id.* at 81-87, esp. 87. (The actual method of cataloguing fingerprints was devised by Edward Henry, a British Police official in India in the early 1890s ("…with Henry's new classification system, a search in a file of more than 8,000 cards required less than five minutes.") *Id.*
[443] See, *Id.* at 81-87, esp. 87. ("…with Henry's new classification system, a search in a file of more than 8,000 cards required less than five minutes.") *Id.*
[444] An example of this is that based on the testimony of a single government witness, a U.S. District Court judge in *U.S. vs Havvard*, 117 F. Supp. 2d 848 (SD IN 2000), stated at 855, ".. the court believes that latent print identification is the very archetype of reliable expert testimony."
[445] See, COLE, *supra* note 1, at 81-87.
[446] See e.g., *U.S. v Llera Plaza*, No. 98-362-10 (E.D.Pa. 01/07/2002); 2002.EPA.0000003 ¶¶ 89-91 <http://www.versuslaw.com> ("The determination that a fingerprint examiner makes . . . when comparing a latent fingerprint with a known fingerprint, specifically the determination that there is sufficient basis for an absolute identification is not a scientific determination. It is a subjective determination standard. It is a subjective determination without objective standards to it. *Id.* at ¶¶ 89-91 (Test. Stoney, Tr. July 12, 1999, at 87). Dr. Stoney's point that "[t]he determination that a fingerprint examiner makes . . . when comparing a latent fingerprint with a known fingerprint . . . is a subjective determination," was fully confirmed by the testimony presented by government witnesses Ashbaugh and Meagher.") *Id.* at ¶ 83.
[447] Francis Galton, *Finger Prints* (Macmillan and Co., London 1892), 65-66. Quoted in COLE, *supra* note 1, at 77.
[448] *Id.*
[449] Cole, *supra* note 1, at 177 (conviction), 202-203 ("pseudo-science"). In that case, Thomas Jennings was convicted of a Chicago murder based solely on fingerprint evidence. After his conviction was affirmed on appeal he was hanged by the State of Illinois on February 16, 1912. *Id.* at 177.
[450] *Id.* at 201-202, 284.
[451] See e.g., "Scientific Methods," *supra* note 24, at 119-120 (Among the requirements of the scientific method are: "testable consequences;" "repetition of the test;" and "reliability and accuracy."). *Id.*
[452] Alan L. McRoberts, "Fingerprints: What They Can & Cannot Do!," *SCAFO*, Vol. 10 Issue 6 (June 1994), at 1-7, http://www.scafo.org/library/100601.html (last visited September 21, 2004).
[453] Dr. Stoney is Director of the McCrone Research Institute in Chicago. Dr. Stoney is quoted in Cole, *supra* note 1, at 284.

[454] *U.S. v Llera Plaza*, No. 98-362-10 (E.D.Pa. 01/07/2002); 2002.EPA.0000003 ¶ 100 <http://www.versuslaw.com> (quoting from, *U.S. v. Mitchell*, CR No. 96-407 (E.D. PA), Dr. Stoney, Test. Stoney, Tr. July 12, 1999, at 87.).

[455] Cole, *supra* note 1, at 202-203 (Fingerprint analysis was compared to "pseudo-sciences" such as palmistry.). See also, (the British Home Office rejected the use of fingerprints for identification purposes in 1894, because "there was no reason to resort to an unproven technology like fingerprints."). *Id.* at 81.

[456] *Id.* at 177 (conviction), 202-203 ("pseudo-science"). In that case, Thomas Jennings was convicted of a Chicago murder based solely on fingerprint evidence. After his conviction was affirmed on appeal he was hanged by the State of Illinois on February 16, 1912. *Id.* at 177.

[457] *Id.* at 202-203 ("pseudo-science"). On February 14, 2002 the Federal Trade Commission filed a complaint against Miss Cleo for deceptive practices. See, Press Release, "FTC Charges "Miss Cleo" Promoters with Deceptive Advertising, Billing and Collection Practices," *Federal Trade Commission*, February 14, 2002. See at, http://www.ftc.gov/opa/2002/02/accessresource.htm (last visited September 18, 2004). The civil accusation's against Miss Cleo for obtaining money through deception were similar in principle to the criminal charges pending against forensic technician Fred Zain at the time of his death in December 2002. Those charges were four counts of obtaining his salary and witness fees under false pretenses by lying on the witness stand and faking test results. "Death of Lying Chemist Fred Zain," *supra* note 181.

[458] Lee, *supra* note 431, at 321.

[459] See e.g., Cole, *supra* note 1, at 201-202, 284. (Fingerprint analysis is a form of 'black art' dependent on the subjective opinion of each examiner, and does not involve any objective and duplicatable standard.)

[460] *The Oxford English Dictionary*, OED Online, defines 'black art' as "magic" in Def. 1. http://dictionary.oed.com (last visited September 18, 2004).

[461] See e.g., Cole, *supra* note 1, at 171. (There is a world of difference between an inked fingerprint, taken carefully and methodically on a clean smooth surface at a police station, and a latent fingerprint, left by accidental contact on some irregular dirty surface at the scene of a crime.) *Id.*

[462] See e.g., Woodworth, *supra* note 6, at esp. 50 ("Fingerprints, you will note, are almost never set down at some crime scene under any BUT poor conditions or in any other way than extremely carelessly."); See also, Cole, *supra* note 1, at 177 (Discusses that early claims of the accuracy of fingerprint matching was based on having a whole fingerprint sample, and not a partial print that is common at crime scenes.) *Id.* The impact such factors can have on a fingerprint sample is emphasized by the possibility that two fingerprint samples taken from a person under perfect laboratory conditions may not be "matched" as being from the same person. See e.g., Lee, *supra* note 431, at 281-283, (Discussing the problems associated with getting accurate fingerprint impressions using *electronic* scanning devices. These include inconsistent contact; non-uniform contact; irreproducible contact; feature extraction artifacts; and sensing. *Id.* It is interesting that Henry Faulds believed that fingerprints could accurately be matched if both prints were obtained under controlled conditions. Cole, *supra* note 1, at 174. However he was adamantly opposed to their use for forensic purposes because of the high likelihood a latent crime scene print could be falsely matched to a person it didn't originate from. *Id.* at 173-176 In this regard he wrote in 1905, "The ordinary rules of evidence require to be even more diligently and methodically employed in the case of so delicate a method, which officials not scientifically trained are apt to misunderstand or overstate in their natural eagerness to secure convictions. "Repeat patterns" in single fingers are often found which come so near, the one to the other, that the least smudginess in the printing of them might easily veil important divergences in one or two lineations, which appalling results. I can

Endnotes   173 hardly emphasize this point too strongly." *Id.* 176, quoting Henry Faulds, *Guide to Finger-Print Identification* (Wood Mitchell (Publishers) 1905), p. 51.

[463] See e.g., McRoberts, *supra* note 452, at 1-7. See e.g., Woodworth, *supra* note 6, at esp. 50-51. (Includes a discussion of various factors related to the two variables of the fingertip's ridges and the quality of the surface that can affect the quality of a fingerprint impression.).

[464] See e.g., Woodworth, *supra* note 6, at 51.

[465] Steve Berry, "Pointing a Finger at Prints," *Los Angeles Times*, Feb 26, 2002, at A1 ("The FBI estimates [latent prints] are one-fifth the size of the inked prints defendants give at booking.") *Id.*

[466] See e.g., *Id.* at A1. ("The FBI estimates [latent prints] are one-fifth the size of the inked prints defendants give at booking.") *Id.*

[467] See e.g., Woodworth, *supra* note 6, at esp. 49-50; See also, Cole, *supra* note 1, at 177. (Discusses that early claims of the accuracy of fingerprint matching was based on having a whole fingerprint sample, and not a partial print that is common at crime scenes.) The impact such factors can have on a fingerprint sample is emphasized by the possibility that two fingerprint samples taken from a person under perfect laboratory conditions may not be "matched" as being from the same person. See e.g., LEE, *supra* note 431, at 281-283, in which it is discussed the problems associated with getting accurate fingerprint impressions using *electronic* scanning devices. These include inconsistent contact; non-uniform contact; irreproducible contact; feature extraction artifacts; and sensing. *Id.*

[468] See e.g., Woodworth, *supra* note 6, at esp. 50. (Fred Woodworth, a professional printer, explains this phenomena by comparing the imprinting of a fingerprint on a surface to the printing of a document on a letterpress. He elaborated on how using the same metal printing plate (letterpress) under controlled conditions in a printing plant will result in documents that cannot be matched as having the same origin by simply varying the pressure of the impression as the ink is applied to the paper.

[469] *U.S. v. Parks*, CR-91-358-JS (C.D. CA), trial transcript, vol. 5 (Dec. 11, 1991), 538-556, referred to in Cole, *supra* note 1, at 272.

[470] *Random House Webster's Unabridged Dictionary* (1999), *latent*: 1. present but not visible, apparent, or actualized; existing as potential.

[471] See e.g., McRoberts, *supra* note 452, at 1-7.

[472] See e.g., *Id.* at 1-7.

[473] *Id.* at 1-7 (In the sense of being invisible and needing a form of interpretation by an "expert" to be understandable to non-experts, fingerprint and DNA evidence are similar.)

[474] See e.g., Woodworth, *supra* note 6, at 47, referring to "Fingering Fingerprints," *The Economist*, December 14, 2000, Science and Technology section. The article explains that fingerprints, the touchstone of forensic science, have never been subjected to proper scientific scrutiny. ("And two other things make the situation worse in practice. The first is that fingerprints found at crime scenes tend to be incomplete. What are being compared are not whole prints, but mere fragments... The second difficulty is that most fingerprint evidence found at the scene of a crime is 'latent'. In other words, it requires treatment ... to make it visible enough to work with - and even then, it is often indistinct. How valid it is to compare such 'filtered' evidence with the clean crisp prints obtained from suspects in controlled conditions is another unexplored question...") *Id.*

[475] See e.g., *Id.*

[476] See e.g., Woodworth, *supra* note 6, at 50. Drawing on decades of experience as a printer dealing with issues related to variances in the quality of an impression, Fred Woodworth explains that it can be caused by the medium *being impressed* (such as a mug at a crime scene), the medium *leaving the impression* (such as a person's perspiration mixed with various contaminants), the medium *applying the impression* (such as a person's elastic fingertips), and

the pressure applied by the originating medium to the receptive medium to *make the impression* (such as the angle and force applied by a person's fingertips on a mug). Woodworth explains that only varying the pressure with which an imprint is made can alters its appearance to such a degree that two items printed from the same originating source can look as if they had different sources. Woodworth raises the proposition that the same principles applying to variations in the quality of a printed item apply to the circumstances that can affect the quality of a fingerprint impression and subsequent identifications. *Id.* Deformation can also occur in fingerprints obtained in a controlled environment. See e.g. Lee, *supra* note 431, at.283. ("Rolled inked fingerprints tend to have a large area of valid ridges and furrows, but have *large deformations due to the inherent nature of the rolled acquisition process.*") (emphasis added). *Id.*

[477] See e.g., McRoberts, *supra* note 452, at 1-7.

[478] See e.g., Berry, *supra* note 465, at A1 ("The FBI estimates [latent prints] are one-fifth the size of the inked prints defendants give at booking.") *Id.*

[479] See e.g., Woodworth, *supra* note 6, at 51.

[480] *Id.* at 51.

[481] *Id.* at 51.

[482] For an explanation of this idea, see, *Id.* at 51. This article also appeared in another publication edited by Fred Woodworth, *The Mystery and Adventure Series Review*, No. 34, Summer 2001.

[483] *Id.* at 51. To make this excerpt as relevant as possible for the reader, "this article" was substituted for the original phrase, "the present issue of The Match," which is a magazine published by Fred Woodworth. This article also appeared in Woodworth, *The Mystery and Adventure Series Review*, No. 34, Summer 2001.

[484] The effect of computer enhancement on the reliability of fingerprint evidence is discussed later in this section. *Infra* notes 590-600 and accompanying text.

[485] Woodworth, *supra* note 6, at 51.

[486] Cole, *supra* note 1, at 171.

[487] See e.g., *Id.* at 177. ("Probabilistic arguments like Balthazard's [circa 1910], however, were aimed at determining the likelihood of whole single fingerprints matching exactly in every particular. They completely overlooked the question relevant to forensic identification, which entailed matching partial fingerprint fragments. This was perhaps too subtle an argument for defense barristers to mount at the time.") *Id.*

[488] *Id.* at 177.

[489] *Id.* at 177.

[490] The partial latent print is compared with a partial area of a suspects print, or if a search is performed, with a partial area of a multitude of people's fingerprints.

[491] See e.g., *Id.* at 89. ("The detective must match this distorted crime scene print to an inked print, taken under pristine "laboratory" conditions, to the exclusion of all other fingerprints in the world.") *Id.*

[492] *Supra* notes 478-485 and accompanying text.

[493] Cole, *supra* note 1, at 177. This is also true to varying degrees in other countries that accepted fingerprint testimony as scientific evidence of a culprit's identity.

[494] *Id.* at 177. A contemporaneous account of the Jennings case is: L. H. L., "The "Finger-Print" Case, Michigan Law Review, Vol. 10, No. 5 (March 1912) pp. 396-401. Online at, http://www.jstor.org/stable/1274485.

[495] *Id.* at 177.

[496] *People v Jennings* 96 N.E. 1077, 252 Ill. 534 (Ill. 1911) (Trial transcript 137-139). Cole, *supra* note 1, at 179.

[497] *Id.*

[498] Cole, *supra* note 1, quote at 179.
[499] *Id.* at 178.
[500] *Id.* at 178.
[501] *Id.* at 177-180. (See the account of the Jennings case).
[502] Quoted in *Id.* at 181. Quote attributed to: Albert Wehde and John Nicholas Beffel, *Finger-Prints Can Be Forged* (Chicago: The Tremonia Publishing Company, 1924), 49-50.
[503] Cole, *supra* note 1, at 179.
[504] *Jennings*, 96 N.E. at 1082-83. See also, Cole, *supra* note 1, quote at 179.
[505] *Id.* at 180, the author described the ground breaking impact of the case in the following way, "Thus the Jennings case established both the admissibility of fingerprint evidence and the exclusive of "experts" to testify for it."
[506] *Id.* The court's approved the admittance of fingerprint evidence "on the basis of general agreement [by experts] and anecdotal evidence. ..." *Id.* At that time an expert was considered as someone with "greater study respecting certain subjects" such as fingerprinting, than a layperson. *Id.*
[507] The author's research did not result in the discovery of a single instance since the *Jennings* case of a jury that did not convict a defendant after the prosecution's expert witness(es) identified his or her fingerprint as matching that of a latent crime related print. See e.g., Haber, *supra* note 60, at 14. ("Jurors believe fingerprint evidence is true. Illsey (1987) showed that jurors place great weight on the testimony of fingerprint experts, and rank fingerprint evidence as the most important scientific reason why they vote for conviction. Meagher (2002) could not remember an instance in which an FBI examiner made a positive identification and the jury set the identification aside and acquitted the defendant.") *Id.*
[508] Cole, *supra* note 1, at 177.
[509] There was a dearth of circumstantial evidence of Thomas Jenning's guilt, and absent prejudice against him because of his skin color and the "expert" fingerprint testimony, there is little reason to believe he would have been found guilty. Because of the substantive doubts about Jenning's guilt, he is listed in, *The Innocents Database, supra* note 3.
[510] See e.g., Cole, *supra* note 1, at 201, 261 and 272 (Judge Letts said in *U.S. v. Parks*, CR-91-358-JS (C.D. CA), *supra*, "... so far as I can tell, department policy is to be comfortable with whatever you have to be in order to get below the number [of] comparisons available.") *Id.*
[511] It is significant that Bertillon *faked* the prints he presented as matching. Cole, *supra* note 1, at 201. ("Bertillon published two *different* fingerprints which ostensibly showed sixteen matching points of similarity." (emphasis in original)). *Id.* Bertillon's demonstration showed "it was conceivable that two different prints showing sixteen points of similarity could exist – in artifice, if not in nature." *Id.* at 201. The British Home Office reiterated the importance of the 16 point standard in 1953. *Id.* at 260. "In 1978, the British National Conference of Fingerprint Experts had voted overwhelmingly to retain the sixteen-point standard." *Id.* at 270.

In our digital age when people long dead can be made to appear in movies (see e.g., Forest Gump), and singers can be made to appear to be engaging in a duet with a long dead performer (e.g., Natalie Cole singing in a duet of *Unforgettable* with her deceased father Nat King Cole), it is childs play for the matching of two dissimilar fingerprints to be digitally faked. The pathway to doing that is eased by the FBI's digitization of the tens of millions of fingerprints in its fingerprint database. Thus there is no technological impediment to the *faking* of a match between a crime scene print and the print of a suspect. Any print can be made to appear to match any other print irrespective of any actual similarities. The ease of doing that is emphasized by how easily entire photographs of people and places can be faked. See e.g., David Kushner, "These Are Definitely Not Scully's Breasts: Inside One Man's Crusade To Save Gillian Anderson and the rest of the world from the plague of fake celebrity

Menace To The Innocent

porn," *Wired Magazine*, November 2003, at 142-145. (Article discusses the faking of celebrity photos, and that in some cases they are so skillfully forged that it is impossible to detect it is a fake.). Furthermore, sophisticated digital techniques are not required to fake a photograph – it can be done with standard printing techniques. See e.g., Woodworth, *supra* note 6, at 51. ("Can a fingerprint be reproduced? Of course it can -- at left, below, you see such a reproduction, a print, in ink, on paper. A printing plate, carefully enough made, can easily contain far more detail than necessary to duplicate the print from a human finger. Compared to the amount of detail in a halftone plate for printing a photograph, for instance, where the plate must hold information on about 40,000 dots per square inch, the reproduction of a fingerprint is a relatively trivial matter.") *Id.*

[512] Cole, *supra* note 1, at 202 ("... a latent print comparison resulting in fewer than sixteen points of similarity would be declared inconclusive automatically.") *Id.*

[513] *Id.* at 201-202.

[514] *Id.* at 201-202.

[515] Shannon P. Duffy, "Experts May No Longer Testify That Fingerprints 'Match'," *The Legal Intelligencer*, January 9, 2002, available at http://www.lawgenie.com/FingEXP.htm (last visited September 20, 2004).

[516] The absence of a minimum meant that for all practical purposes the FBI's minimum was at least one, since if there were none then there would be no basis for a "match." However beginning in 1983, a school of thought has developed that fingerprints can be matched from their *ridges* without relying on any perceived *points* of commonality. See e.g., COLE, *supra* note 1, at 266-269. Known as ridgeology, it is based on the idea that there are no points on a fingertip, only ridges. *Id.* at 267. Thus, "When all friction ridge formations present are in agreement and the examiner is of the opinion that there are sufficient unique details present to eliminate anyone else as a possible donor, the examiner has formed an opinion of identification." *Id.* at 267. The inventor of ridgeology, Canadian Mounted Policeman David Ashbaugh, contends it makes "it possible to effect identification with fewer points than most agencies required, as few as three or four." *Id.* at 267. That is, ridgeology's supporters claim it can be used to match a sample (latent) fingerprint to a control fingerprint that may only have three or four matching points using the point method. *Id.* at 267. However that claim needs to be viewed from the perspective that ridgeology suffers from the same lack of accuracy verification as the point method, and both are subject to questions concerning the uniqueness of fingerprints. *Id.* at 268. As Simon Cole, the author of *Suspect Identities* observed, "After all, knowing how ridges are formed does not actually prove they are unique, nor, if we simply assume they are unique in some absolute sense, does it measure *how similar* different friction ridge arrangements might be." *Id.* at 268.

[517] *U.S. v. Parks*, No. CR-91-358-JSL (C.D. CA). Quote in Cole, *supra* note 1, at 273.

[518] The aura of fingerprint uniqueness gained a solid foothold because of an absence of a scientific challenge to the idea, as fingerprint evidence became more widely used as an identification method. See e.g., Cole, *supra* note 1, at 180, 187-189.

[519] *U.S. v. Llera Plaza*, 188 F.Supp.2d 549, 569 (E.D.Pa. 03/13/2002). See also, Cole, *supra* note 1, at 263.

[520] *Id.* at 263.

[521] *Id.* at 205. ("correspondence schools [were] "turning out thousands of so-called 'University' graduates with the degree of F.P.E."'). *Id.*

[522] The book was co-authored by journalist John Beffel and fingerprint forger Albert Wehde. See, *Id.* at 188, 276-278.

[523] *Id.* at 188, 276-277.

[524] *Id.* at 276-277.

[525] *Id.* at 276-277.

[526] *Id.* at 277-278, citing note 25 at 342. (note 25 cites, IAI, *Proceedings* (1927), 39-43). The potential danger to their livelihood expressed by fingerprint examiners in 1927, three years after Wehde's book appeared, seemed to at least in part vindicate his portrayal of "fingerprint examiners as self-interested, obligated, by interest in preserving their own livelihood, to defend the validity of fingerprint identification against all challenges." *Id.* at 188.
[527] *Id.* at 188, 274-278.
[528] *Id.* at 275.
[529] *Id.* at 275.
[530] *Id.* at 275. (The police chief, August Vollmer said of Brown's technique, "It will take much more than the mere discovery of means of transferring the mark left by a criminal to tear down the great fingerprint system which, even though young, has assumed immense proportions.")
[531] *Id.* at 275.
[532] *Id.* at 275.
[533] *Id.* at 275.
[534] *Id.* at 275. (emphasis in original)
[535] *Id.* at 275. (emphasis in original)
[536] "Handwriting Expert Has Found a Method of Copying Impressions," *Oakland Tribune*, September 13, 1913, p. 2. Kytka claimed forged fingerprint evidence could used to "railroad" and convict innocent persons of crimes they didn't commit. It was described that "The world universally accepts the infallibility of the finger-print identification."
[537] Bertillon's demonstration influenced England to adapt a minimum standard of 16 common points between a latent print and a sample before a match could be declared. See, Cole, *supra* note 1, at 201. ("Bertillon published two *different* fingerprints which ostensibly showed sixteen matching points of similarity." (emphasis in original)), See also, *Id.* at 260, 270.
[538] Even the respected Henry C. Lee ignores this important point in a 2001 book he co-edited, in which he acknowledged Bertillon's body measurement system. See, Lee, *supra* note 431, at 276.
[539] Cole, *supra* note 1, at 2. ("French police official Alphonse Bertillon's anthropometric system of identification used the lengths of bones to track individualized identities.")
[540] Quote in, *Id.* at 32.
[541] See e.g., *Id.* at 201 ("Bertillon's argument had been made: it was conceivable that two different prints showing sixteen points of similarity could exist – in artifice, if not in nature.")
[542] Richard Austin Freeman, *The Red Thumb Mark* (London: House of Stratus 2001) at 111 (Originally published in the United Kingdom 1907). The major character was a doctor and lawyer named John Thorndyke. R. Austin Freeman's character of Dr. Thorndyke could be considered a literary rival of his contemporary Sir Arthur Conan Doyle's Sherlock Holmes. Both main character of both authors used scientific methods of deduction and logic to solve seemingly insoluble problems. The popularity of Freeman's books in his own day is reflected in the fact that 97 years after its initial publishing date, *The Red Thumb Mark* is still in print.
[543] *Id.* at 111. *The Merriam-Websters Dictionary*, 10th Collegiate Ed. defines Syllogism, 1: a deductive scheme of a formal argument consisting of a major and a minor premise and a conclusion (as in "every virtue is laudable; kindness is a virtue; therefore kindness is laudable.")
[544] Cole, *supra* note 1, at 274-278.
[545] *Id.* at 278.
[546] *Id.* at 278.
[547] *Id.* at 278-279.
[548] *Id.* at 274.
[549] *Id.* at 274, 280.
[550] *Id.* at 274.

[551] *Id.* at 280.

[552] *Id.* at 274, 280.

[553] *Id.* at 279-280. Quoting from, Nelson E. Roth, "The New York State Police Evidence Tampering Investigation: Confidential Report To The Honorable George Pataki Governor Of The State Of New York," (Albany, NY: New York State Police, Jan. 20, 1997), at 110.

[554] Cole, *supra* note 1, at 279-280. Roth, *supra* note 553, at 110.

[555] Cole, *supra* note 1, at 280. Roth, *supra* note 553, at 312.

[556] Cole, *supra* note 1, at 280.

[557] See e.g., a discussion in this regard in, *Id.* at 279-281.

[558] *Id.* at 279.

[559] *Id.* at 274.

[560] *Id.* at 274.

[561] *Id.* at 274.

[562] The known cases of crime lab malfeasance and/or outright manufacturing or doctoring of evidence, indicates that either accidental discovery or a whistleblower is how such cases are brought to the public's attention. An example is Fredric Whitehurst's whistleblowing of irregularities in the FBI's crime lab. See *supra* notes 111-130 and accompanying text.

[563] Clifford Irving, *Fake! the Story of Elmyr de Hory the Greatest Art Forger of Our Time*, (New York: McGraw-Hill 1969). (This biography of the greatest art forger of the twentieth century is written by Clifford Irving, who is best remembered for his forgery of the autobiography of Howard Hughes.). *F for Fake* (1973, ASIN: 6303473261), a documentary about Elmyr de Hory produced and directed by Orson Welles, incorporated the unique twist of telling the story of a great art forger in France, that he revealed at the end of the film was a fake forger he dreamed up to illustrate how easily people can be deceived about what is real and what is simply a false representation of what is real.

[564] Cole, *supra* note 1, at 275.

[565] Freeman, *supra* note 542.

[566] The ease of creating fake fingerprints is emphasized by the known use of digitization to fake photographs to the degree that in the absence of compelling documentary evidence to the contrary, a fake photograph is indistinguishable from being original. See e.g., *Kushner*, supra note 511. ("It's mind-boggling what he [a master faker] can do."). *Id.* at 145. A fingerprints level of detail is minuscule compared to that of a photograph. Woodworth, *supra* note 6, at 51.

[567] *Id.*

[568] In 1997 an investigator for a major U.S. insurance company related to the author that using off the shelf computer software, scam artists were creating the paperwork necessary to support a fraudulent insurance claim by manufacturing all the necessary police reports, witness statements, hospital records, doctors reports, prescription orders, etc. The sophistication of commercially available digital imagery software has increased since 1997.

[569] The ease of creating fake fingerprints is emphasized by the known use of digitization to fake photographs to the degree that in the absence of compelling documentary evidence to the contrary, a fake photograph is indistinguishable from being original. See e.g., *Kushner*, supra note 511. ("It's mind-boggling what he [a master faker] can do.") *Id.* at 145. A fingerprints level of detail is minuscule compared to that of a photograph. Woodworth, *supra* note 6, at 51.

[570] For a discussion of various issues related to computerization of fingerprints, see, Lee, *supra* note 431, at 281-282, 319, 383; and, Pankanti, *supra* note 430.

[571] The FBI began digitally scanning fingerprint images in 1972, and in 1979 it began testing an automated fingerprint search system (AFIS). By 1980 14.3 million scanned fingerprint cards were in the FBI's computer database. Cole, *supra* note 1, at 252-253.

[572] Pankanti, *supra* note 430, at 3, 5-6.

[573] *Id.* at 5.

[574] For a discussion of issues related to an AFIS' determination that prints are "sufficiently similar," see, Pankanti, *supra* note 430, at 3, 5-6.

[575] An indicator that the AFIS error rate is significant is indicated by the fact that a manufacturer of biometric fingerprint equipment estimated the FRR (false reject rate) of its device was .03%. That estimate was found to understate the devices' error rate by a factor of over *83,333%* when, "An independent test by the Sandia National Laboratory found that the same system had an FRR of 25%..." Lee, *supra* note 431, at 319.

[576] A number of the problems with computer identifications are discussed in, LEE, *supra* note 431, at 281-284.

[577] *Id.* at 281.

[578] Id. at 281-2.

[579] *Id.* at 282.

[580] Id. at 282.

[581] *Id.* at 282.

[582] See e.g., *Id.* at 282.

[583] *Id.* at 282. The minutiae referred to are details of the fingerprint.

[584] *Id.* at 282.

[585] *Id.* at 282.

[586] *Id.* at 282.

[587] *Id.* at 282.

[588] *Id.* at 282.

[589] *Id.* at 282.

[590] *Id.* at 282.

[591] *Id.* at 309. The latent print does not necessarily have to be computer enhanced. For example, if the computer enhancement of a scanned print is printed out on a high-quality printer, it can be visually compared with a latent print by standard techniques.

[592] McRoberts, *supra* note 452, at 1-7. (Due to their invisibility, a latent print must be treated with "some kind of powder, chemical, or electronic processing, or enhancement," to transform it into being visible.)

[593] See e.g., *Fingerprint Evidence*, Nolo Press, at, http://www.nolo.com/lawcenter/ency/article.cfm/objectID/6BC35DBC-8021-41FD-842843A353454430/catID/950198C5-C82B-447A-85A6254081097CFE (last visited Sept. 20, 2004). ("The visibility of a set of prints depends on the surface from which they're lifted; however, with the help of computer enhancement techniques that can extrapolate a complete pattern from mere fragments, and laser technology that can read otherwise invisible markings, fingerprint experts increasingly can retrieve identifiable prints from most surfaces.") *Id.*

[594] Berry, *supra* note 465, at A1. ("The FBI estimates [latent prints] are one-fifth the size of the inked prints defendants give at booking.") *Id.*

[595] Nolo Press, a major publisher of legal books, maintains a webpage on *Fingerprint Evidence*, on which it says, "The visibility of a set of prints depends on the surface from which they're lifted; however, with the help of computer enhancement techniques that can extrapolate a complete pattern from mere fragments, and laser technology that can read otherwise invisible markings, fingerprint experts increasingly can retrieve identifiable prints from most surfaces." *Fingerprint Evidence*, *supra* note 593.

[596] Lee, *supra* note 431, at 309. ("Some methods can estimate the orientation and/or frequency of ridges in each block in the fingerprint image and adaptively tune the filter characteristics to match the ridge characteristics.") An as yet unexplored problem with computerized enhancement techniques, is that the software's coding determines the algorithm used to enhance the print – thus certain fingerprint features can be inordinately enhanced to increase the evidentiary value of the newly created print. In other words, an enhanced print is

Menace To The Innocent

inevitably skewed to some degree by the software's algorithm technique that performs the actual enhancement of the print's features.

[597] According to the FBI, a latent print is typically 1/5th the size of a rolled print it is compared to. Berry, *supra* note 465, at A1. ("The FBI estimates [latent prints] are one-fifth the size of the inked prints defendants give at booking.") *Id.*

[598] Lee, *supra* note 431, at 312.

[599] *Id.* at 309.

[600] *Id.* at 309.

[601] *Id.* at 309.

[602] *Id.* at 309.

[603] *Random House Webster's Unabridged Dictionary, supra* note 470, defines *"unique:* 1. existing as the only one or as the sole example; single; solitary in type or characteristics. 4. limited to a single outcome or result; without alternative possibilities."

The dichotomy of the claim that computer software can accurately reconstruct the missing portions of a fingerprint when "no two people have the same pattern of friction ridges," is emphasized by both claims are made by proponents of fingerprinting as a discriminating identification technique. See e.g., *Fingerprint Evidence, supra* note 593.

[604] Fingerprint enhancement and restoration software posses an incalculably grave threat to innocent people, because in a particular case a suspect's fingerprint in a law enforcement computer database can be used as the template to reconstruct the latent print to match that of the suspect.

[605] See e.g., Chris Brislawn, "The FBI Fingerprint Image Compression Standard," *Chris Brislawn's Website* (One of the designers of the FBI's national standard for wavelet-based compression (WSQ) of their fingerprint database.), http://www.di-srv.unisa.it/~ads/corso-security/www/CORSO-9900/biometria/FBI.htm (last visited September 18, 2004).

[606] PKWARE introduced the PKZIP compression standard in 1989. PKWARE, http://www.pkware.com (last visited September 21, 2004).

[607] See e.g., "Lossless Compression," *Webopedia* (on-line computer term encyclopedia) at, http://www.pcwebopaedia.com/TERM/L/lossless_compression.html (last visited September 20, 2004). ("The PKZIP compression technology is an example of lossless compression.") *Id.*

[608] Brislawn, *supra* note 605.

[609] *Id.*

[610] *Id.* For an explanation of a terabyte of data, see, http://wombat.doc.ic.ac.uk/foldoc/foldoc.cgi?terabyte

[611] Brislawn, *supra* note 605.

[612] *Id.*

[613] *Id.*

[614] *Id.*

[615] See e.g., "Lossless Compression," *supra* note 607. ("[Lossy] refers to data compression techniques in which some amount of data is lost.") *Id.*

[616] Brislawn, *supra* note 605. (Discusses the "distortion in the compressed images" caused by loosy techniques.)

[617] *Id.*

[618] Lee, *supra* note 431, at 284.

[619] Brislawn, *supra* note 605.

[620] *Id.* (A sample fingerprint image was compressed by JPEG to 45,853 bytes, and WSQ compressed it to 45,621 bytes – a difference of approximately ½ of 1%.)

[621] *Id.* ("The [FBI's] standard is a discrete wavelet transform-based algorithm referred to as Wavelet/Scalar Quantization (WSQ).") *Id.*

[622] See e.g., "Lossless Compression," *supra* note 607. ("[Lossy] refers to data compression

techniques in which some amount of data is lost.") *Id.*

[623] *Supra* notes 478-485 and accompanying text, esp. ("…it is likely that multiple sections of every person's fingers are identical at some scale of comparison, to a section of any number of other people's fingers.") Woodworth, *supra* note 6, at 51.

[624] See e.g., "Lossless Compression," *supra* note 607. ("[Lossy] refers to data compression techniques in which some amount of data is lost.") *Id.*

[625] The effect of computer enhancement on the reliability of fingerprint evidence was discussed previously in this section. See, *supra* notes 591-602 and accompanying text. The effect of scanning a fingerprint on the reliability of fingerprint evidence was discussed previously in this section. See *supra* notes 571-590 and accompanying text.

[626] *Id.*

[627] Kelly, *supra* note 25, at 16. (During an interview with author Phillip Wearne in March 1997, a former forensic lab technician was brutally frank in observing, "People say we're tainted for the prosecution. Hell, that's what we do! We get our evidence and present it for the prosecution.") *Id.*

[628] A personal example of this is the author had occasion to not use computerized spell check for a period of time, and a noticeable result was a sharpening of the author's spelling skills. This benefit was lost after the use of a computer's spell check feature was once again available.

[629] See, Haber, *supra* note 60, at 8. ("For a forensic science, it is of critical importance to identify and understand the sources of errors.") *Id.*

[630] *Daubert v. Merrell Dow Pharmaceuticals,* 509 U.S. 579 (1993). *Daubert* established a four-pronged test for a federal court to use in considering the reliability of otherwise relevant expert testimony considered for admission under under *Fed. R. Evi.* 702: four factors–testing, peer review, error rates, and "acceptability" in the relevant scientific community–which might prove helpful in determining the reliability of a particular scientific theory or technique, *id.,* at 593—594. See also, Cole, *supra* note 1, at 284. The defense motion challenging the scientific veracity of expert fingerprint testimony was that based on the *Daubert* standard, it didn't meet the admissibility standard of expert testimony concerning scientific evidence. Under *Frye v. United States,* 293 F. 1013 (D.C. Cir. 1923), the standard replaced by *Daubert* for admissibility under *Fed. R. Evi.* 702 was merely criteria four: "general acceptance in the particular field in which it belongs." *Id* at 1014. In *Kumho Tire Co. v. Carmichael,* 526 U.S. 135 (1999), the Court ruled, "(a) The Daubert "gatekeeping" obligation applies not only to "scientific" testimony, but to all expert testimony. Rule 702 does not distinguish between "scientific" knowledge and "technical" or "other specialized" knowledge, but makes clear that any such knowledge might become the subject of expert testimony." 526 U.S. at 138. The *Kumho* case arose due to the attempts of prosecutors and civil litigants to circumvent *Daubert* by presenting testimony as being based on specialized or technical – not "expert" – knowledge.

[631] See, *U.S. v Mitchell,* 199 F.Supp 2d 262, 2002.EPA.0000039 ¶ 12 < http://www.versuslaw.com> (Mitchell was granted a retrial after his first trial resulted in his conviction. See, *U.S. v. Mitchell,* 145 F.3d 572 (3rd Cir.1998).

[632] *Daubert,* 509 U.S. 579 (1993)

[633] *U.S. v Llera Plaza,* No. 98-362-10 (E.D.Pa. 01/07/2002); 2002.EPA.0000003 ¶ 165 <http://www.versuslaw.com> (The FBI crime lab sent prints to 53 crime labs, of which 34 responded.); For a news account, see, Andy Newman, "Fingerprinting's Reliability Draws Growing Court Challenges," *The New York Times,* April 7, 2001.

[634] *Id.* at ¶ 165. (Prints were sent to 53 crime labs, of which 34 responded.); For a news account, see, Newman, *supra* note 633.

[635] *Id.* at ¶ 165. ("Nine of the thirty-four responding agencies did not make an identification in

the first instance."). The breakdown of the results from those nine was, "Seven laboratories failed to match one of the latent prints with Mitchell's inked prints, and five failed to match the second latent print." Brian Carnell, *The Myth of Fingerprints*, April 10, 2001, available at, Skepticism.net, http://www.skepticism.net/articles/2001/000004.html. (last visited September 20, 2004). For a news account, see, Newman, *supra* note 633.

[636] *U.S. v Llera Plaza*, No. 98-362-10 (E.D.Pa. 01/07/2002); 2002.EPA.0000003 ¶ 165 <http://www.versuslaw.com>.

[637] *Id.* at ¶ 165.

[638] *Id.* at ¶ 165.

[639] *Id.* at ¶ 165.

[640] *Id.* at ¶ 165.

[641] *Id.* at ¶ 165.

[642] *Id.* at ¶ 165.

[643] *Id.* at ¶ 165.

[644] *Id.* at ¶ 268 note 23.

[645] *Id.* at ¶ 268 note 23.

[646] *Id.* at ¶ 268 note 23.

[647] *Id.* at ¶ 268 note 23. ("...in a training status, if an individual fails to make an identification that we believe they should have been able to, we would take that information back to that individual, show them the characteristics of which they should take into consideration, ask them to reassess their position and, you know, use the information that's now presented to them and try to come up with the same conclusion. That is, that the two prints were identical.") *Id.*

[648] *U.S. v. Mitchell*, CR No. 96-407 (E.D. PA)

[649] *Supra* notes 60-89 and accompanying text discussing the ASCLD fingerprint proficiency tests administered from 1983 to 1991, the CTS proficiency tests from 1995 to 2001, and the AIA's proficiency tests from 1993-2001 (2001 is the most current year that results are available). It will also be noted that the minimum false positive rate of the initial *Mitchell* test was 26.5% if the defendant's print and the latent crime scene prints actually *did* match, while if the prints *didn't* match the false positive rate would have been the inverse, or 73.5%. *Id.*

[650] The various proficiency tests that fingerprint examiners have taken are described in, Haber, *supra* note 60, at esp. 5-11, and 16-19.

[651] See e.g., LEE, *supra* note 431, at 383.. ("From a statistical viewpoint, the scientific foundation for fingerprint individuality is incredibly weak.") Fingerprint individuality is still being seriously questioned in the 21st century, and yet it is the basis of expert fingerprint testimony – that the defendant is *the source* of the latent crime related fingerprint. It is also important to keep in mind that errors can be in the form of a false-positive or a false-negative, although if for no reason other than the pro-prosecution bias exhibited by crime lab technicians, there is reason to believe the former are more common than the latter.

[652] *U.S. v Llera Plaza*, No. 98-362-10 (E.D.Pa. 01/07/2002); 2002.EPA.0000003 ¶ 146 <http://www.versuslaw.com>. (emphasis added to original) ("Q: Based on the uniqueness of fingerprints, friction ridge, etcetera, do you have an opinion as to what the error rate is for the work that you do, latent print examinations? A: As applied to the scientific methodology, it's zero." *U.S. v. Mitchell*, CR No. 96-407 (E.D. PA)), Testimony of FBI technician Meagher, Tr. July 8, 1999, at 154-56.

[653] Newman, *supra* note 633.

[654] *U.S. v Llera Plaza*, No. 98-362-10 (E.D.Pa. 01/07/2002); 2002.EPA.0000003 ¶ 201 <http://www.versuslaw.com>.

[655] *U.S. v. Llera Plaza*, 188 F.Supp.2d at 575-576.

[656] *Id.* at 576.

[657] *Id.* at 575-576. If the British had not abandoned the 16-point standard six months earlier (June 11, 2001), the basis for Judge Pollak's reversal wouldn't have existed. *Id.* at 569.

[658] Cole, *supra* note 1, at 272-273 (Citing *U.S. v. Parks*, CR-91-358-JSL (C.D. CA).

[659] *Id.* at 272-273 (Citing *U.S. v. Parks*, CR-91-358-JSL (C.D. CA). (Judge Letts excluded the fingerprint testimony under the much less stringent *Frye* standard of admissibility under Fed. R. Evi. 702.)

[660] *Id.* at 272-273 (Citing *U.S. v. Parks*, CR-91-358-JSL (C.D. CA).

[661] *Id.* at 272 (Citing *U.S. v. Parks*, CR-91-358-JSL (C.D. CA).

[662] *Id.* at 273. (Citing *U.S. v. Parks*, CR-91-358-JSL (C.D. CA), trial transcript, vol. 5 (Dec. 11, 1991), 585-607).

[663] *Id.* at 273 (Citing *U.S. v. Parks*, CR-91-358-JSL (C.D. CA), trial transcript, vol. 5 (Dec. 11, 1991), 585-607).

[664] In a 2002 case U.S. District Judge Louis Pollak initially granted a defense motion to bar testimony that "in the opinion of the witness, a particular latent print is - or is not - the print of a particular person." *U.S. v Llera Plaza*, No. 98-362-10 (E.D.Pa. 01/07/2002); 2002.EPA.0000003 ¶ 201 <http://www.versuslaw.com>. Upon reconsideration at the prosecution's request, Judge Pollak reversed himself and ruled testimony would be allowed as to the opinion of an expert examiner for either party that the defendant's print did or did not match latent crime scene prints. See, *Llera Plaza*, 188 F.Supp.2d at 576.

[665] Hans Sherrer, "It's About Time! Fred Woodworth's Questioning of Fingerprint Evidence Is Long Overdue," *Justice Denied*, Vol. 2, Issue 8, p. 10, at, http://www.justicedenied.org/fredwoodwortharticles.htm (last visited September 16, 2004). See also, John Wesley Noble and Bernard Averbuch, *Never Plead Guilty: The Story of Jake Ehrlich* (Farrar, Straus and Cudahy 1955), 295-298.

[666] *Id.*

[667] *Id.*

[668] James E. Starrs, "Saltimbancos on the Loose? Fingerprint Experts Caught in a Whorl of Error," *The Scientific Sleuthing Newsletter*, Vol. 12, No. 2, Spring 1988, at 1, 5.

[669] *Id.* at 1, 5.

[670] *Id.* at 5.

[671] *Id.* at 5.

[672] *Id.* at 5. The exclusion of Basden as the fingerprints source by enlargement of the latent prints is eerily similar to an incident in *The Riddle of the Stone Elephant* (Grosset & Dunlap) a 1949 mystery by Sam Epstein. In that book two prints that looked similar were only able to be declared as dissimilar by a police fingerprint expert after being blown up to two feet in height. See a discussion of this episode in the book in, Fred Woodworth, "The Opening Salvo," *Justice Denied*, Vol. 2, Issue 9, p. 46, at, http://www.justicedenied.org/openingwoodworth.htm (Last visited September 16, 2004).

[673] Starrs, *supra* note 668, at 5.

[674] *Id.* quote at 5. The exclusion of Basden as the fingerprints source by enlargement of the latent prints is eerily similar to an incident in *The Riddle of the Stone Elephant* (Grosset & Dunlap) a 1949 mystery by Sam Epstein. In that book two prints that looked similar were only able to be declared as dissimilar by a police fingerprint expert after being blown up to two feet in height. See a discussion of this episode in the book in, Woodworth, *supra* note 672, at 46.

[675] Starrs, *supra* note 668, quote at 5.

[676] *Id.* at 6.

[677] *Id.* at 6. (The rapist was named the "prime time rapist" because he attacked women as they were watching nighttime television programs.)

[678] *Id.* at 6.

[679] *Minnesota v Roger Sipe Caldwell*, 322 N.W. 2d 574, 580 (1982). The right thumb print

and the print on the envelope were testified to have matched at 11 points. Berry, *supra* note 465, at A1.

[680] *Id.* at 579.

[681] *Id.* at 580.

[682] *Id.* at 582, 586.

[683] *Id.* at 582, 584, 586.

[684] *Id.* at 582. There was no conclusive evidence presented by the prosecution that Roger Caldwell travelled from Golden, CO to Duluth, Minnesota between the hours of 10pm and 2am. Their contention that depended on concealing the alibi witness, whose testimony was indirectly supported by concealment of the exculpatory thumb print negative, was that Caldwell was in Duluth that evening. *Id.* The distance between the two cities is listed on Indo.com at, http://www.indo.com/ (last visited September 20, 2004).

[685] *Id.* at 584.

[686] *Id.* at 586.

[687] Berry, *supra* note 465, at A1.

[688] *Id.*, at A1.

[689] Ralph Vigoda, "Fingerprints Put To Test," *Philadelphia Enquirer*, January 28, 2003, at, http://www.philly.com/mld/inquirer/news/front/5045177.htmhttp://www.philly.com/mld/inqui rer/news/front/5045177.htm (last visited September 18, 2004).

[690] *Id.*

[691] *Id.*

[692] *Id.*

[693] *Id.*.

[694] *Id.*

[695] *Stephan Cowans*, The Innocence Project, http://www.innocenceproject.org/case/display_profile.php?id=141. (September 18, 2004).

[696] *Id.*

[697] *Id.*

[698] *Id.*

[699] *Id.*

[700] *Id.* (In contrast, the people closest to the assailant and most likely to reliably identify him, did not finger Cowans.).

[701] *Id.*

[702] *Id.*

[703] *Id.*

[704] *Id.* Conflicting with that was the identification two weeks after the shooting by the policeman, who may have been traumatized by the event, of Cowans out of an array of eight photographs. On that same day a person looking out of a second-floor window also identified Cowans. In contrast, the people closest to the assailant and most likely to reliably identify him, did not finger Cowans. *Id.*

[705] Les Zaitz, "FBI Case Against Oregon Lawyer Built on Blurry Fingerprint, Logic," *The Oregonian*, May 30, 2004, http://www.oregonlive.com/news/oregonian/index.ssf?/base/front_page/10858322779300.xml (last visited October 2, 2004). (Brandon Mayfield was arrested as a material witness.)

[706] *Id.*

[707] *Id.*

[708] Affidavit of Rickard K. Werder, May 6, 2004, *In Re: Federal Grand Jury* 03-01, No. 04-MC-9701 (USDC WD OR), ¶ 7.

[709] *Id.* at ¶ 7.

[710] *Id.* at ¶ 8.

711 Zaitz, *supra* note 705.
712 Les Zaitz, "Transcripts Detail Objections, Early Signs of Flaws," *The Oregonian*, May 26, 2004.
713 Zaitz, *supra* note 705.
714 *Id.*
715 David Heath, "FBI's Handling of Fingerprint Case Criticized," *The Seattle Times*, June 1, 2004.
716 *Id.*
717 *Id.*
718 *Id.*
719 Sarah Kershaw, "Spain and U.S. at Odds on Mistaken Terror Arrest," *The New York Times*, National Section, June 5, 2004.
720 Zaitz, *supra* note 705.
721 Noelle Crombie and Les Zaitz, "FBI Apologizes to Mayfield," *The Oregonian*, May 25, 2004.
722 Andrew Kramer, "Brandon Mayfield Opens Investigations Into His Arrest," *KATU 2 News* [Portland, OR], May 25, 2004.
723 Mayfield's lawsuit is pending as of December 2004.
724 Blaine Harden, "FBI Faulted in Arrest of Oregon Lawyer," *Washington Post*, November 16, 2004, A02. (Report published in the *Journal of Forensic Identification*, November-December, 2004).
725 *Id.*
726 *Id.*
727 *Id.*
728 *Id.* (Mayfield's lawyer, Steven Wax, asked in response to the November 2004 report, "Is this a fingerprint mistake, or does it go beyond the business of fingerprinting with the contamination of the process with other information?"). By the wording of his question, Wax indicates an unawareness that since the fingerprinting process is subjective, it is susceptible to any number of influences unrelated to making an identification based on the fingerprint image under examination. See e.g., *U.S. v Llera Plaza*, No. 98-362-10 (E.D.Pa. 01/07/2002); 2002.EPA.0000003 ¶¶ 89-91 <http://www.versuslaw.com> (Quoting, Test. Stoney, Tr. July 12, 1999, at 87).
729 On June 11, 2001 England adopted the FBI's policy of not requiring a minimum number of matches before two fingerprint samples can be declared as uniquely similar. See, U.S. v. *Llera Plaza*, 188 F.Supp.2d at 568-569.
730 Cole, *supra* note 1, at 282. Seven horses were also killed in the blast.
731 "Danny McNamee," *Innocent: Fighting Miscarriages of Justice*, http://www.innocent.org.uk/cases/dannymcnamee/index.html (last visited Sept. 18, 2004). See also, Cole, *supra* note 1, at 282.
732 "McNamee," *supra* note 730.
733 *Id.*
734 Cole, *supra* note 1, at 282.
735 *Id.* at 282.
736 *Supra* notes 502-512 and accompanying text.
737 *U.S. v. Llera Plaza*, 188 F.Supp.2d at 569.
738 *Id.* at 569. See also ("the ACE-V process employed by New Scotland Yard is essentially indistinguishable from the FBI's ACE-V process"), *Id.* at 575-576. Cole, *supra* note 1, at 286. (Britain abandoned the sixteen-point standard in 2001.)
739 McNamee's conviction was quashed in December 1998, 30 months before abandonment of the 16 point standard on June 11, 2001. Since there were no short-cuts taken in the McNamee

case, it indicates that other innocent people in Britain since June 2001, and in this country for decades, have been wrongly convicted based on faulty fingerprint examinations, and they have been unable to have that error corrected on appeal as McNamee was able to do. *McNamee, supra* note 730.

[740] *Id.* See also, Cole, *supra* note 1, at 282.

[741] Among these tactics is the false linking of a person with an unpopular group. In Danny McNamee's case, the prosecution falsely linked him with the IRA, to which he did not belong and wasn't associated. *Danny McNamee, supra* note 730.

[742] See e.g., Editorial ("The FBI's flawed lab"), *supra* note 2. (The FBI crime lab engages in "shoddy work, withholding of relevant evidence from defense attorneys and outright bias in favor of prosecutions."); see also, (During an interview with author Phillip Wearne in March 1997, a former forensic lab technician was brutally frank in observing, "People say we're tainted for the prosecution. Hell, that's what we do! We get our evidence and present it for the prosecution.") Kelly, *supra* note 25, at 16.

[743] For general discussions of the prevalence of prosecutors participating in the framing of a suspect, see, Sherrer, *supra* note 34; see also, *Prosecutorial Lawlessness, supra* note 34.

[744] See e.g., Cole, *supra* note 1, at 206. By pleading guilty to avoid what may be an unavoidable wrongful conviction, the innocent person avoids further suffering by having the "trial penalty" imposed on him or her, which typically means that a person convicted after a trial is sentenced to a longer term than someone who pleads guilty.

[745] David Brand, "Fingerprint Evidence," *Cornell University News*, January 24, 2002, at http://www.news.cornell.edu/releases/Jan02/fingerprint.study.deb.html (last visited September 16, 2004).

[746] See e.g., Kurland, *supra* note 23, at 153. (Speech by then FBI Director William Sessions on September 1, 1998.) DNA is short for Deoxyribonucleic Acid.

[747] See e.g., *Id.* at 153.

[748] *Id.* at 153 (Sessions speech was on September 1, 1988. (emphasis added)).

[749] See e.g., Thompson, *supra* note 10. Considered as a whole, the article undermines the concept that DNA analysis provides scientifically certain evidence of a DNA sample's exact source. See Kelly, *supra* note 25, at 231-232.

[750] Kelly, *supra* note 25, at 231-232.

[751] *Id.* at 231-232.

[752] Dan L. Burk, "DNA Identification: Possibilities and Pitfalls Revisited," 31 *Jurimetrics* 53 (Fall 1990), at 80.

[753] Kelly, *supra* note 25, at 231-232.

[754] The prevalence of DNA false positives underscores the impossibility of establishing a person's guilt with nothing more than a DNA test result. See, e.g., Thompson, *supra* note 10. This is a modern day confirmation of author R. Austin Freeman's common sense admonition in his 1907 book, *The Red Thumb Mark*, "But there is no such thing as a single fact that 'affords evidence requiring no corroboration.' As well might one expect to make a syllogism with a single premise." Freeman, *supra* note 542, at 111.

[755] See e.g., Cole, *supra* note 1, at 71. Author Simon Cole is specifically referring to the subjectivity of fingerprint analysis, but his observations apply with equal force to DNA analysis. See also, Baden, *supra* note 46, at 231-232 for a discussion of the subjective element involved in evaluation of scientific tests ("Four hundred years ago, Sir Francis Bacon entitled his seminal work on the subject of science Novum Organum. Loosely translated form the Latin, it means "a new instrument," in this case, a new instrument of reasoning. Bacon wrote: "The human understanding is not a dry light, but is infused by desire and emotion, which give rise to a 'wishful science.' For man prefers to believe what he want to be true. He therefore rejects difficulties, being impatient of inquiry; sober things, because they restrict his hope;

deeper parts of nature, because of his superstition; the light of experience, because of his arrogance and pride, lest his mind should seem to concern itself with things mean and transitory; things that are strange and contrary to all expectation, because of common opinion. In short, emotion in numerous, often imperceptible ways pervades and infects the understanding." … Others have furthered our understanding of the impact of "wishful science" on the pursuit of pure science." Nobel Prize winning chemist Irving Langmuir coined the term pathological science for "the science of things that aren't so," and Richard Feynman, a Nobel winner for his work in physics, focused on the source of the trouble when he warned, "The first principle is that you must not fool yourself and you're the easiest to fool." Much as I hate to admit it, the sad fact is that some forensic scientists do, indeed, fool a lot of the people a lot of the time.") *Id.* at 231-232.

[756] See e.g., Thompson, *supra* note 10, at 47. ("A false positive might occur due to error in the collection or handling of samples, misinterpretation of test results, or incorrect reporting of test results.")

[757] Ruth Teichroeb, "Call for a review of state crime labs," *Seattle Post-Intelligencer*, September 14, 2004, B1, B5 (The Washington state crime lab system is officially known as the Forensic Laboratory Services Bureau.).

[758] *Id.*

[759] *Id*

[760] Matthew R. Durose and Patrick A. Langan, Ph.D., "State Court Sentencing of Convicted Felons, 2000 – Table 4.1 Estimated number of felong convictions in State courts," *Bureau of Jutsice Statistics*, U. S. Dept. of Justice, June 2003, NCJ 198822, at https://www.bjs.gov/content/pub/pdf/scscf00.pdf (last viewed September 16, 2004)

[761] See e.g., Thompson, *supra* note 10, at 52. DNA's use as confirmatory evidence is consistent with the fact that, "A false positive might occur due to error in the collection or handling of samples, misinterpretation of test results, or incorrect reporting of test results." *Id.*

[762] Press Release, "DOJ Aware of Problems in FBI's DNA Lab," *NACDL*, November 25, 1997, available at, https://www.nacdl.org/newsreleases.aspx?id=18127 (last visited September 19, 2004)

[763] For example, in November 1994 the California Supreme Court ruled for the first time that testimony related to DNA evidence alone was sufficient to sustain a conviction – in the absence of any other testimonial or physical evidence. See, *People v. Soto*, 48 Cal.App.4th 924, 30 Cal.App.4th 340, 34 Cal.App.4th 1588, 39 Cal.App.4th 757, 43 Cal.App.4th 1783, 35 Cal.Rptr.2d 846; 1994.CA.40950 <http://www.versuslaw.com> (Cal.App. Dist.4 11/22/1994) Frank Lee Soto had been accused of raping a 79 year-old who was unable to speak due to a stroke. In its majority opinion the Court stated, "Because DNA RFLP is so highly reliable and relevant, "to allow a minor academic debate … to snowball to the point that it threatens to undermine the use of it in court" is throwing the baby out with the bath water." *Id.* at ¶55. However the as yet all but ignored evidence related to the prevalence of false-positives, and the multitude of reasons underlying them, is an indication that courts acted hastily by placing total faith in crime lab related DNA testimony as a method of establishing guilt without *any corroborating* evidence.

[764] DNA evidence is circumstantial due to its confirmatory nature, in contrast with conclusatory evidence such as a video tape showing a particular person robbing a bank. See e.g., Press Release ("DOJ Aware"), *supra* note 762 ("…there are no certainties in science, only probabilities.") *Id.*

[765] DNA testing has actually been likened to "DNA Fingerprinting." See e.g., M. Krawczak and J. Schmidtke, "DNA Fingerprinting," *BIOS Scientific Publishers*, 1998).

[766] This commonality is extends to the newer identification method being referred to as DNA fingerprinting. See e.g., "DNA Fingerprinting," *Encyclopædia Britannica* Online,

http://search.eb.com/eb/article?eu=31233 (last viewed September 19, 2004).

[767] Thompson, *supra* note 10, at 47.

[768] *Id.* at 47.

[769] *Id.* at 47.

[770] See, *Schlup v. Delo*, 513 U.S. 298, 324-325 (1995).

[771] The O.J. Simpson trial showed that the burden can be overcome. In that case it was accomplished by a combination of of painting a plausible scenario that incriminating blood (DNA) evidence could have been planted, and that L.A. crime lab conditions and handling procedures made contamination of DNA samples a likely possibility – thus skewering any test results. See e.g., Kelly, *supra* note 25, at 250.

[772] Thompson, *supra* note 10, at 47.

[773] *Id.* at 47.

[774] See e.g., *Id.* at 47.

[775] Josiah Sutton is one such person falsely convicted on the basis of an erroneous DNA analysis and trial testimony. See, Adam Liptak, "You Think DNA Evidence Is Foolproof. Think Again," *The New York Times*, March 16, 2003.

[776] *Id.*

[777] Press Release ("UCI Professor"), *supra* note 374.

[778] Roma Khanna and Steve McVicker, "New DNA test casts doubt on man's 1999 rape conviction," *Houston Chronicle*, March 10, 2003, at www.chron.com/cs/CDA/ssistory.mpl/special/crimelab/1812821 (last visited September 21, 2004).

[779] Press Release ("UCI Professor"), *supra* note 374.

[780] *Id.*

[781] *Id.*

[782] *Id.*

[783] Olsen, *supra* note 375.

[784] *Id.*

[785] *Id.*

[786] *Schlup v. Delo*, 513 U.S. at 324-325 ("Indeed, concern about the injustice that results from the conviction of an innocent person has long been at the core of our criminal justice system. That concern is reflected, for example, in the "fundamental value determination of our society that it is far worse to convict an innocent man than to let a guilty man go free."") (citation omitted) See also, T. Starkie, *Evidence* 756 (1824) ("The maxim of the law is . . . that it is better that ninety-nine . . . offenders shall escape than that one innocent man be condemned."), *Schlup* 513 U.S. at 325.

[787] Press Release ("UCI Professor"), *supra* note 374.

[788] Thompson also found numerous other cases involving irregularities. See, *Id.*

[789] Liptak, *supra* note 775. On June 27, 2003 the Harris County prosecutor announced he was going to recommend an Executive Pardon for Sutton. *Id.*

[790] See e.g., Kurland, *supra* note 23, at 153 (Speech by former FBI Director William Sessions about the certainty of DNA evidence.).

[791] Connors, *supra* note 169, at 34.

[792] *Id.*, at 34.

[793] See e.g., Thompson, *supra* note 10, at esp. 47-48.

[794] See e.g., *Id.* at esp. 47-48.

[795] See e.g., *Id.* at esp. 47. ("By contrast, no court has rejected DNA evidence for lack of valid, scientifically accepted data on the probability of a false positive. ... but [it is considered] unnecessary to have comparable estimates on the frequency of false positives."). *Id.*

[796] See e.g., Kelly, *supra* note 25, at 250. (The DNA of 225 FBI agents taken 14 months apart under perfect laboratory conditions was falsely matched by FBI lab technicians at the rate of

12-1/2% – 1 out of 8.)

[797] *Id.* at 250.

[798] *Id.* at 250.

[799] *Id.* at 250.

[800] Thompson, *supra* note 10, at 48.

[801] Among the factors that can cause a false positive are errors "in sample handling, procedure, or interpretation." National Research Council, *DNA Technology in Forensic Science*, (National Academies Press 1992), at 88.

[802] Thompson, *supra* note 10, at 47.

[803] National, *supra* note 801, at 88.

[804] *Id.* at 88-89.

[805] J. Koehler, "DNA Matches and Statistics," 76 *Judicature* 222, 229 (1993). See also, J. Koehler, "Error and Exaggeration in the Presentation of DNA Evidence at Trial," 34 *Jerimetics* 21, 24 (1993) [hereinafter, Koehler 2] (false positive error rate of 1-4% shown by proficiency testing). Both articles cited in *People v. Marshall*, No. BA-069796 (Sup Ct. LA County) (Motion To Exclude DNA Evidence, October 10, 1995).

[806] Richard Lempert, "After the DNA Wars:skirmishing with NRC II," *Jurimetrics* 37: 439-468 (1997).

[807] Koehler, *supra* note 805, at 229. See also, Koehler 2, *supra* note 805, at 24 (1993) (false positive error rate of 1-4% shown by proficiency testing). *Id.* Both articles cited in *People v. Marshall*, No. BA-069796 (Sup Ct. LA County) (Motion To Exclude DNA Evidence, October 10, 1995).

[808] For a litany of crime lab conditions and practices that escalate the probability of a false positive, see *supra* Part 2: Shoddy Work is the Norm for Crime Labs.

[809] See e.g., Underwood, *supra* note 4, at 167 ("It is no secret that expert witnesses can be "co-opted" by the prosecution-they may be little more than hired guns of the state."), referencing note 87 that cites, William C. Thompson, "A Sociological Perspective on the Science of Forensic DNA Testing," 30 *U.C. Davis L. Rev.* 1113, 1115 (1997).

[810] Thompson, *supra* note 10, at 48. This is also true in civil related DNA testing, such as one to establish paternity.

[811] Alok Jha, "DNA Fingerprinting 'No Longer Foolproof'," *The Guardian* (UK London), September 9, 2004, at http://www.guardian.co.uk/uk_news/story/0,3604,1300068,00.html (last visited September 29, 2004)

[812] *Id.*

[813] *Id.*

[814] "NDIS Statistics," *Combined DNA Index System*, http://www.fbi.gov/hq/lab/codis/clickmap.htm (last visited September 29, 2004).

[815] Dr. Bruce Lipton has written and lectured extensively about this phenomena and the findings of the Human Genome Project See e.g., an interview of Dr. Lipton on April 17, 2001, and other the information available at, Bruce Lipton, Ph.D., "Conversation for Exploration," *lauralee.com*, http://www.lauralee.com/lipton.htm (last visited September 19, 2004). One thing Dr. Lipton has reported on is that a person with multiple personalities can have physical characteristics and the underlying DNA change in seconds, such as their eye color changing from blue in one personality to brown in another. *Id.* See also, Barry Commoner, "Unraveling the DNA Myth: The Spurious Foundation of Genetic Engineering," *Harper's Magazine*, February 2002, at 39-47, "… alternative splicing can be said to generate *new* genetic information." (emphasis in original) *Id.* at 43; and, "…most molecular biologists operate under the assumption that DNA is the secret of life, whereas the careful observation of the hierarchy of living processes strongly suggests that it is the other way around: DNA did not create life, life created DNA. … DNA is a mechanism created by the cell to store

Menace To The Innocent

information produced by the cell." *Id.* at 47. Article available at: http://www.commondreams.org/views02/0209-01.htm (last visited September 19, 2004). Thus one's DNA changes to reflect the change in one's cellular structure, which accounts for the changes attributable to a person with multiple personalities. Lipton, *supra* note 815.

[816] Dr. Bruce Lipton has written and lectured extensively about this phenomena and the findings of the Human Genome Project See e.g., an interview of Dr. Lipton on April 17, 2001, and other the information available at, *Bruce Lipton, Ph.D.*, *supra* note 815. One thing Lipton has reported on is that a person with multiple personalities can have physical characteristics and the underlying DNA change in seconds, such as their eye color changing from blue in one personality to brown in another. See also, Commoner, *supra* note 815, "... alternative splicing can be said to generate *new* genetic information." (emphasis in original) *Id.* at 43; "...most molecular biologists operate under the assumption that DNA is the secret of life, whereas the careful observation of the hierarchy of living processes strongly suggests that it is the other way around: DNA did not create life, life created DNA. ... DNA is a mechanism created by the cell to store information produced by the cell." *Id.* at 47. Thus one's DNA changes to reflect the change in one's cellular structure, which accounts for the changes attributable to a person with multiple personalities. Lipton, *supra* note 815.

[817] The author does not know the "random match probability" that was testified to at Sutton's trial. However a billion to one is being used to demonstrate that even with a high probability that a coincidental match is unlikely, the false positive probability can still be so low as to not meet the 99% probability of guilt suggested by the Supreme Court as necessary to sustain a finding of guilt. See, *Schlup*, 513 U.S. at 324-325.

[818] This is on the high end, since it was recently noted, "commentators have suggested, the rate of false positives is between 1 in 100 and 1 in 1000, or even less ...", Thompson, *supra* note 10, at 49.

[819] Unless of course the jurors wanted a free lunch or dinner or simply wanted to prolong Josiah Sutton's agony of uncertainty before pronouncing him guilty.

[820] For an explanation of this phenomena, see e.g., "Prosecutorial Lawlessness," *supra* note 34. ("The pervasive lawlessness of prosecutors who will resort to any tactic in their desperation *to win at all costs* was documented in the exposé and reports published by various publications, including the *Pittsburgh Post-Gazette*, the *Chicago Tribune*, the, *Reason* magazine, and Amnesty International.") *Id.*

[821] Thompson, *supra* note 10, at 49. See also, Krawczak, *supra* note 765.

[822] See e.g., Thompson, *supra* note 10, at 50. Table 1: Posterior odds that a suspect is the source of a sample that reportedly has a matching DNA profile, as a function of the relation between prior odds, random match probability, and false positive probability.

[823] See e.g., *Id.* at 50. Table 1: Posterior odds that a suspect is the source of a sample that reportedly has a matching DNA profile, as a function of the relation between prior odds, random match probability, and false positive probability.

[824] Indeed, with a random match probability of 1 in 1,000 the odds would have been more likely that Sutton *was not* the source of the DNA sample, than that he was. See, *Id.* at esp. 51.

[825] See, *Schlup,* 513 U.S. 324-325. Furthermore, the odds in Sutton's case only matched the 10% probability of innocence that is sufficent to warrant acquittal inferred in Blackstone's admonition, "it is better that ten guilty persons escape, than that one innocent suffer." Thompson, *supra* note 10, at 51.

[826] Press Release ("UCI Professor"), *supra* note 374.

[827] Prosecutors use smoke and mirror tactics to bolster weak or non-existent evidence in many types of cases. For an analysis of how they are used to obtain convictions in tax related cases, see, Sherrer, *supra* note 314, at 102-122.

[828] See e.g., Thompson, *supra* note 10, at 47. ("[N]o court has rejected DNA evidence for lack

of valid, scientifically accepted data on the probability of a false positive.")
[829] *Id.* at esp. 47.
[830] *Id.* at 47.
[831] *Id.* at 48.
[832] *Id.* at 48.
[833] *Id.*; See also, William C. Thompson, "Accepting Lower Standards, The National Research Council's second report on forensic DNA evidence," *Jurimetrics* 1997;37(4):405-424.
[834] David Feige, "The Dark Side of Innocence," *The New York Times*, June 15, 2003.
[835] See, The Innocents Database, *supra* note 3. There have also been people exonerated by DNA evidence in Australia and the U.K. As of September 2004, The Innocents Database included over 2,500 people who had been either judicially exonerated or pardoned, based on DNA and other forms of exoneration.
[836] In an effort to obtain reliable statistics concerning the number of people whose conviction significantly relied on DNA evidence that an expert testified was incriminatory, the author contacted several state crime laboratories, the FBI crime laboratory, the office of several state prosecutors, national prosecution organizations, and the Bureau of Justice Statistics (BJS). It appears that as of September 2004 this is a statistic that is not tracked on a state or federal level. However, for purposes of this paper the extraordinarily conservative figure of 10,000 convictions that significantly relied on DNA evidence will be used. The actual number is assuredly much higher, considering that figure is only about 1% of the 984,000 convictions in state and federal court in 2000 alone. See, "Courts and Sentencing Statistics," *Bureau of Justice Statistics* website at http://www.ojp.usdoj.gov/bjs/stssent.htm#selected (last visited March 22, 2004). The 911,842 convictions in 1994, for example, is over 90% of the number of 2000 convictions, and indicates that a figure of 10,000 is likely to be less than 1/10th of 1% (1 out of a 1,000) of the convictions since DNA began to be used in US courts. See e.g., Patrick A. Langan and Jodi M. Brown, *Felony Sentences in the United States, 1994*, Bureau of Justice Statistics, U. S. Dept. of Justice, Revised Sept. 17, 1997, NCJ 165149, https://www.bjs.gov/content/pub/pdf/fsus94.pdf (last visited Sept. 20, 2004). Furthermore, the reliance of prosecutors on DNA testimony is growing, particularly in more populous jurisdictions, which likewise have more prosecutions. For example, the BJS reported, "In 2001 *two-thirds* of prosecutors' offices reported the use of DNA evidence during plea negotiations or felony trials. This is an increase from 1996 when about half offices reported using DNA evidence during plea negotiations or felony trials, *98% of full-time medium offices*, 73% of full-time small offices, and 38% of part-time offices." Carol J. DeFrances, "Prosecutors in State Courts, 2001," *Bureau of Justice Statistics, U. S. Dept. of Justice*, May 2002, NCJ 193441, quoted at page 8, https://www.bjs.gov/index.cfm?ty=pbdetail&iid=1125 (last visited Sept. 20, 2004) (emphasis added).
[837] That estimate is supported by a detailed analysis that over 14 percent of all convictions in state and federal courts are of innocent people. See, Sherrer ("How Many"), *supra* note 215. The conservativeness of this estimate is indicated by the fact that it was found by a study of all 4,578 capital appeals finalized in the U.S. between 1973 and 1995, that "7% of capital cases nationwide are reversed because the condemned person was found to be innocent." The study was overseen by Columbia University School of Law Professor James Liebman, who was co-author of the final report. See, Liebman, *supra* note 215. A summary of the report is: Hans Sherrer, "Landmark Study Shows the Unreliability of Capital Trial Verdicts," *Justice Denied*, Vol. 2, No. 2 (Nov. 2000), available at, http://justicedenied.org/landmarkstudy.htm (last visited September 27, 2004).
[838] See, Thompson ("How the Probability"), *supra* note 10 at 48.
[839] Connors, *supra* note 169, at 34-35.
[840] *Id.* at 48.

Menace To The Innocent

[841] *Id.* at 48.
[842] *Id.* at 48.
[843] *Id.* at 48.
[844] *Id.* at 48. After Gerald Davis' retrial, the indictment was dismissed against his father Dewey Davis, who was his co-defendant and who also spent eight years wrongly imprisoned.
[845] *Id.* at 56. ("Police serologist Fred Zain ... testified that the genetic markers in the semen left by the assailant matched those of Harris and only 5.9 percent of the population.")
[846] *Id.* at 57.
[847] *Id.* at 57.
[848] *Id.* at 57. (William Harris spent seven years in prison and one year of home confinement while waiting to have his conviction vacated.)
[849] For numerous examples of institutionalized forensic lab/examiner abuse, see *supra* Chapter 2: Shoddy Work is the Norm for Crime Labs.
[850] Lefcourt, *supra* note 2.
[851] *Id.*
[852] *Id.* (In contrast with a blind test, the technicians had the advantage of knowing they were being tested, and they were aware of the parameters of the test.).
[853] *Id.* (Alan Robillard, head of the lab's DNA Unit in 1989, acknowleged to investigators for the Office of the Inspector General "that he ordered the results of the proficiency test destroyed, claiming the test was flawed." The test was supposedly readministered, with *all* the technicans passing. However that claim is suspect because the FBI has not produced any records concerning the alleged retest and its results.). *Id.*
[854] See *supra* Chapter 2: Shoddy Work is the Norm for Crime Labs.
[855] This account of Kim Ancona's murder is based on, Hans Sherrer, "Twice Wrongly Convicted of Murder – Ray Krone Is Set Free After 10 Years," *Justice Denied*, Vol. 2, Issue 8 (March 2002), at 25-26, and the sources cited therein. Available at, http://www.justicedenied.org/volume2issue8.htm#Ray (last visited September 16, 2004).
[856] *Id.* at 25-26.
[857] *Id.* at 25-26.
[858] *Id.* at 25-26.
[859] *Id.* at 25-26.
[860] *Id.* at 25-26.
[861] *Id.* at 25.
[862] *Id.* at 25.
[863] *Id.* at 25.
[864] Henry Weinstein, "Death Penalty Foes Mark a Milestone: Arizona convict freed on DNA tests is said to be the 100th condemned U.S. prisoner to be exonerated since executions resumed," *Los Angeles Times*, April 10, 2002.
[865] *Id.*
[866] *Id.* (63.5% false positives + 22% false negatives = 85.5% erroneous results. The 14.5% accurate result is approximately 1 out of 7 tests.)
[867] Sherrer, *supra* note 855.
[868] It is possible that a number of allegedly scientific methods of evidence evaluation would not be able to withstand a well prepared *Daubert* challenge, such as the several successful challenges to admitting handwriting analysis testimony. See e.g., Pankanti, *supra* note 430.
[869] It is possible that a number of allegedly scientific methods of evidence evaluation would not be able to withstand a well prepared *Daubert* challenge, such as the several successful challenges to admitting handwriting analysis testimony. See e.g., *Id.*
[870] Kelly, *supra* note 25, at 277.
[871] *Id.* at 277.

[872] *Id.* at 277.

[873] See e.g., "Scientific Methods," *supra* note 24, at 119-120 (Among the defining characteristics of the scientific method are: "testable consequences;" "repetition of the test;" and "reliability and accuracy.") *Id.*

[874] *Horstman v. Florida*, 530 So.2d 368 (Fla.App.2 Dist 1988).

[875] *Id.* at 369-370.

[876] *Id.* at 370.

[877] *Id.* at 370.

[878] Connors, *supra* note 169, at 54.

[879] *Id.* at 54.

[880] *Id.* at 55.

[881] *Id.* at 55.

[882] *Id.* at 57-59.

[883] *Id.* quote at 58.

[884] *Id.* at quote 58.

[885] *Id.* at 58. ("In the original trial, a forensics expert testified that sperm was present in the semen on the vaginal swab. The prosecutor contended that the sperm was the boyfriend's...")

[886] *Id.* at 59.

[887] *Id.* at 59. (Prosecutor's concurred with Honaker's clemency petition).

[888] Raymond Bonner, "Death Row Inmate Is Freed After DNA Test Clears Him," *The New York Times*, August 24, 2001, A11.

[889] *Id.*

[890] *Id.* (Author note: Although the admissibility of a polygraph examination's results vary from state to state, prosecutors sometimes use them as a scale-tipper to decide against prosecuting a person who receives a favorable report. For example, the author personally knows of someone accused of manslaughter who had charges dropped after a polygraph test examiner agreed that he was telling the truth in denying being at the scene of the person's death, and that he had nothing to do with the death. The test results were corroborated by multiple alibi witnesses who placed him at a party miles away from where the man was killed.)

[891] *Id.*

[892] Connors, *supra* note 169, at 73-74.

[893] *Id.* at 73. (An Alford plea is not an admittance of guilt by a defendant, but it is a public concession that irrespective of the person's innocence, there is enough potentially incriminating (usually) circumstantial evidence that the person is likely to be convicted after a trial.)

[894] *Id.* at 73.

[895] *Id.* at 73-74. The FBI report also said that someone with Vasquez' mental impairments could not have committed the crime, irrespective of the other man's identification as the culprit.

[896] *Id.* at 74.

[897] *Jackson v. Florida*, 511 So. 2d 1047 (Fla.App.2 Dist 1987).

[898] *Id.* at 1048. (A defense expert disputed that Jackson's bite mark matched those found on the victim. 511 So. 2d at 1049).

[899] *Id.* at 1048.

[900] *Id.* at 1049.

[901] *Id.* at 1049.

[902] The Court of Appeal Criminal Division was created in 1907. Prior to then there was no judicial review of a criminal conviction in Britain. A pardon by the Crown was the established way of rectifying an obviously wrongful conviction. See e.g., "England to have a Criminal Court of Appeal," *The New York Times*, April 29, 1906, available at,

http://query.nytimes.com/gst/abstract.html?res=9A06EED9113EE733A2575AC2A9629C946
797D6CF (last viewed September 27, 2004).
[903] This account of the case involving Lord Willoughby and Adolph Beck is related in
Kurland, *supra* note 23, at 115-116.
[904] *Id.* at 115-116.
[905] *Id.* at 115-116.
[906] *Id.* at 115-116.
[907] *Id.* at 115-116.
[908] *Id.* at 115-116.
[909] *Id.* at 115-116.
[910] *Id.* at 115-116.
[911] *Id.* at 115-116.
[912] *Id.* at 115-116.
[913] *Id.* at 115-116.
[914] *Id.* at 115-116.
[915] *Id.* at 116.
[916] Morton, *supra* note 902.
[917] Kurland, *supra* note 23, at 116.
[918] *Id.* at 116.
[919] The Court of Appeal Criminal Division was created in 1907. Prior to then there was no
judicial review of a criminal conviction in Britain. A pardon by the Crown was the established
way of rectifying an obviously wrongful conviction. See e.g., Morton, *supra* note 896.
[920] For a full account of the Herbert Andrews case, see, Edwin M. Borchard, *Convicting the
Innocent: Sixty-Five Actual Errors of Criminal Justice*, (Yale University Press 1932), at 1-6.
[921] *Id.* at 1-6.
[922] *Id.* at 1-6.
[923] *Id.* at 1-6.
[924] *Id.* at 1-6.
[925] *Id.* at 1-6.
[926] *Id.* at 1-6.
[927] *Id.* at 1-6.
[928] *Id.* at 1-6.
[929] *Id.* at 1-6.
[930] *Id.* at 1-6.
[931] *Id.* at 1-6.
[932] *The Winslow Boy*, (Review by Sarah Barnett), culture@home, 2002,
http://www.anglicanmedia.com.au/old/cul/TheWinslowBoy.htm (last visited September 19,
2004).
[933] *Id.*
[934] See e.g., *Id.*
[935] See e.g., Underwood, *supra* note 4, at 177.
[936] *Id.* at 177.
[937] *Alfred Dreyfus, The Innocents Database, supra* note 3, at http://forejustice.org/db/Dreyfus-
-Alfred-.html (last visited September 17, 2004).
[938] The definitive account of this was written by Irving after he was convicted of fraud related
to his attempted caper, and had completed his prison sentence. See, Clifford Irving, *Hoax*
(Danbury CT: Franklin Watts 1981). The story of Irving's hoax broke while Orson Welles was
filming a documentary in France that Irving was appearing in. On the fly Welles incorporated
the breaking story into the film: *F for Fake, supra* note 563.
[939] *Id.* A University of Delaware Library webpage that lists sixteen books related to elaborate,

and in many cases successful fraud schemes involving documents is at, http://www.lib.udel.edu/ud/spec/exhibits/forgery/hoaxes.htm.
[940] *Id.*
[941] *Id.*
[942] *Id.* Clifford Irving was subsequently convicted of fraud related to the hoax and he served time in prison.
[943] See e.g., Stephen Fay, Lewis Chester, and Magnus Linklater, *Hoax: the Inside Story of the Howard Hughes—Clifford Irving Affair* (New York: The Viking Press 1972). (This trio of journalists presents a fascinating account of Clifford Irving's near-successful attempt to have his forgery of the autobiography of Howard Hughes published.)
[944] *The Innocents Database, supra* note 3, has summaries of a number of cases involving innocent people wrongly convicted by a reliance of juries and/or judges either on evidence erroneously believed to be scientific, or on faulty eyewitness identification(s).
[945] See e.g., Pankanti, *supra* note 430, at 3. ("Several courts have now ruled that handwriting identification does not meet the *Daubert* criteria.")
at, http://biometrics.cse.msu.edu/cvpr230.pdf.
[946] *Washington v. Kunze*, 988 P.2d 977 (Wash.App.Div 2 1999).
[947] *Id.* at 979.
[948] *Id.* at 979.
[949] *Id.* at 979.
[950] *Id.* at 980.
[951] *Id.* at 981.
[952] *Id.* at 981.
[953] *Id.* at 981.
[954] *Id.* at 981-987.
[955] *Id.* at 981-983.
[956] *Id.* at 987.
[957] *Id.* at 991.
[958] Associated Press, "Charges Dropped in Earprint Case," March 23, 2001, http://www.truthinjustice.org/mccann.htm (last visited September 16, 2004).
[959] *Id.*
[960] *Id.*
[961] Bob Woffinden, "Earprint Landed Innocent Man in Jail For Murder," *The Guardian* (UK London), January 23, 2004, at http://www.guardian.co.uk/uk_news/story/0,,1129414,00.html (last visited September 16, 2004)
[962] *Id.*
[963] See the editorial in the British Medical Journal's summary of some of the research undermining SBS as a scientifically supportable phenomenon, Editorial, "The Evidence Base For Shaken Baby Syndrome: We need to question the diagnostic criteria," *British Medical Journal*, Vol. 328, March 27, 2004, 719-720. See also, Jackie Dent, "Research Casts Doubt on 'Shaken Baby' Science," *The Guardian*, March 26, 2004, availabe at, http://society.guardian.co.uk/health/story/0,7890,1178919,00.html (last visited September 16, 2004). The two primary studies cited are: Patrick Lantz, S. Stanton, and C. Weaver, "Perimacular Retinal Folds From Childhood Head Trauma: Case report with critical appraisal of current literature," *British Medical Journal*, 2004;328:754-756; and, Mark Donohoe, "Evidence Based Medicine and Shaken Baby Syndrome. Part 1: literature review, 1966-1998," *Am. J. Forensic Med. Pathol.* 2003;24:239-42.
[964] Dent, *supra* note 963. ("British authorities estimate that around 200 babies die from the syndrome each year").
[965] Editorial, *supra* note 963, at 719. See also, Lantz, *supra* note 963, at 754-5. (Researchers at

the Wake Forest University School of Medicine reviewed medical literature and found, "Statements in the medical literature that perimacular retinal folds [bleeding into the eye] are diagnostic of shaken baby syndrome are not supported by objective scientific evidence." *Id.* at 756. The report cites, e.g., the case of a 14-month old child who suffered bleeding into the eye after a television fell on his head. *Id* at 754-5.).

[966] *Id.* at 756.

[967] *Id.* at 754-756.

[968] Dent, *supra* note 963. (The other two being "a certain type of bleeding around the brain, and damage to the brain.").

[969] *Id.* (Dr. Brian Harding of the Great Ormond Street Children's Hospital explaining that bleeding into the eye is used as the singular criteria to diagnose SBS in an unknown number of cases.).

[970] *Id.* ("We need to reconsider the diagnostic critieria, if not the existence, of shaken baby syndrome.")

[971] *Id.* ("We need to reconsider the diagnostic critieria, if not the existence, of shaken baby syndrome.")

[972] Donohoe, *supra* note 963, at 239-42; see also, Dent, *supra* note 934. (Research Donohoe found "the scientific evidence to support a diagnose was much less reliable than generally thought.").

[973] Donohoe, *supra* note 963, at 239-42;

[974] Editorial, *supra* note 963, at 720.

[975] Dent, *supra* note 963. After 21 years of imprisonment Ken Marsh's 1983 conviction of shaking to death his girlfriend's two-year-old son was reversed on August 10, 2004, and charges against him were dropped by the prosecution on September 3, 2004. Evaluation of the child's injuries and maladies excluded him as the source. John Wilkens, "Charges Are Dismissed In 1983 Death," *San Diego Union-Tribune*, September 4, 2004, at http://crossword.uniontrib.com/uniontrib/20040904/news_2m4marsh.html (last visited Sept. 11, 2004). After seven years of imprisonment, Alan Yurko's first-degree murder conviction of shaking to death his ten week old son was vacated on August 27, 2004, after a week-long evidentiary hearing exposed that the child's injuries could have had causes unrelated to parental treatment. See generally, *supra* notes 187-192 and accompanying text.

In September 2004, 38 cases in Britain involving women convicted of killing one of her children were identified after a special review by Attorney General Lord Goldsmith, as possibly bieng unsafe due to "serious disagreement between distinguished and reputable experts," regarding the medical evidence underlying the conviction. Robert Verkaik, "Up to 40 child-killing convictions in doubt," *The Independent*, September 12, 2004, available at: http://www.independent.co.uk/news/uk/crime/up-to-40-child-killing-convictions-in-doubt-7907217.html (last visited September 23, 2004). That was about 14%, one out of seven cases reviewed. *Id.* Twenty-four of the cases were recommended for referal to the Criminal Cases Review Commission (CCRC) and 14 were recommended for appeal. *Id.* Critics suggested the review should be expanded to refer all cases to the CCRC that were based on the testimony of a single expert. *Id.*

Also in September 2004, an 18-month investigation by sponsored by the Royal College of Pathologists and the Royal College of Paediatrics and Child Health resulted in the recommendation of numerous changes to guard against wrongful convictions of a child's "sudden and unexplained death". The report of a working group was chaired by Helena Kennedy QC, "Sudden unexpected death in infancy: A multi-agency protocol for care and investigation," *The Royal College of Pathologists and the Royal College of Paediatrics and Child Health*, September 2004. Available at, https://pdfs.semanticscholar.org/4121/8e74e52cf0b806d53f061383e878f77aa1d3.pdf (last viewed September 23, 2004). The report noted, "Those regularly involved in child abuse can find it hard to be dispassionate and indeed sometimes become hawkish." *Id.* at 4. It also noted that expert witnesses can be "drawn into error because they base their testimony on medical belief rather than scientific evidence." *Id.* at 4.

[976] Connors, *supra* note 169, at xiii.

[977] *United States v. Addison*, 498 F.2d 741, 744 (D.C. Cir. 1974).

[978] The Editors of Encyclopædia Britannica, "DNA Fingerprinting," *The New Encyclopædia Britannica*, 2002 (Rev. 15th Ed.), Vol. 4, pages 140-141 (DNA fingerprinting is "also called DNA Typing, in genetics, method of isolating and making images of sequences of DNA (deoxyribonucleic acid). The technique was developed in 1984 by British geneticist Alec Jeffreys.").

[979] Applebome, *supra* note 20. The effectiveness of this tactic is indicated by the prevalence of innocent people falsely pleading guilty. See e.g, Cole, *supra* note 1, at 206. (Expert fingerprint testimony induces the innocent to plead guilty.). See also, Hans Sherrer, "Medell Banks Jrs.' Conviction for Killing A Non-Existent Child Is Thrown Out As A "Manifest Injustice"," *Justice Denied Magazine*, Vol. 2, Issue 9 (March 2003), pp. 31-35. (Three innocent people pled guilty to murdering a child that never existed when faced with testimony of a doctor that the baby could have existed.).

[980] Huff, *supra* note 34, at 73-74.

[981] Dozens of cases from across the country in which an innocent person falsely pled guilty are listed in, *The Innocents Database*, *supra* note 3.

[982] See e.g., Sherrer, *supra* note 979, at 31-35. (As a ruse to get out of jail on bail the alleged mother, Victoria Banks, feigned being pregnant. A doctor indicated she could be pregnant and she was released a few months before her projected June 1999 due date. Suspicion was raised when no child was seen, and in August 1999 Victoria Banks, her estranged husband Medell Banks Jr., and her sister Dianne Tucker, were interrogated as to the child's whereabouts. All three eventually pled guilty to manslaughter and were sentenced to 15 years in prison when threatened with the death sentence if they were convicted after a trial.)

[983] See e.g., *Id.* at 32-33.

[984] See e.g., *Id.* at 32-33.

[985] For numerous examples of institutionalized forensic lab/examiner abuse, see *supra* Chapter 2: Shoddy Work is the Norm for Crime Labs.

[986] *Id.*

[987] Cole, *supra* note 1, at 270.

[988] *Id.* at 270.

[989] *Id.*

[990] See e.g., *supra* Chapter 3.XIV: Louise Robbins - Forensic Anthropologist; *supra* Chapter 3.XV: Michael West - Forensic Dentist; *supra* Chapter 3.XVI: Sandra Anderson – Cadaver Finding Dog Trainer; and *supra* Chapter 3.XVII: Anthony Pellicano – Audio Expert Extraordinaire.

[991] Kelly, *supra* note 25, at 16. Quoting an interview of the former technician with author Phillip Wearne in March 1997.

[992] See e.g., *Id.* at 16.

[993] Lefcourt, *supra* note 2.

[994] *Id.*

[995] *Id.* (emphasis added to original) (Lab techs are taught how to mislead the judge and jury when, for example, a more sophisticated DNA tests (PCR) exclusion of a suspect contradicts an inconclusive and/or possible inclusory test result of a less sophisticated one (RFLP).).

[996] *Id.* (emphasis in original).

[997] James E. Starrs, "The Ethical Obligation of the Forensic Scientist in the Criminal Justice

System," *Journal of the Association of Official Analytical Chemists*, Vol. 54, No. 4 (1971), pp. 1-12. Quoted in Kelly, *supra* note 25, at 15.
[998] Interview with Phillip Wearne, March 1997, quoted in Kelly, *supra* note 25, at 16. (emphasis added to original).
[999] *Id.* at 15.
[1000] McCloskey, *supra* note 402, at 24. Centurion Ministries is the oldest "innocence project" like organization in the U.S. dedicated to pursuing the exoneration of wrongly convicted people. See, http://www.centurionministries.org (last visited September 27, 2004).
[1001] See e.g., Hans Sherrer, "Introduction to The Innocents, Part One," *Justice Denied*, Vol. 1, Issue 2 (March 1999), ("Contrary to the view it normally projects, what we call the justice system functions as an assembly line designed to produce results in the form of closed files, and not to ensure justice for the men and women who find themselves traveling on its conveyor belt.")
[1002] Cole, *supra* note 1, at 297. (Describes a case involving alleged DNA evidence in which it came to light that the prosecution's expert technicians looked at the evidence *expecting* to see matches, rather than evaluating it dispassionately. DNA is a cousin in principle to fingerprint evidence and uses some the same terminology (see e.g., Krawczak, *supra* note 764.
[1003] Underwood, *supra* note 4, at 167 ("It is no secret that expert witnesses can be "co-opted" by the prosecution-they may be little more than hired guns of the state."), referencing note 87 that cites William C. Thompson, "A Sociological Perspective on the Science of Forensic DNA Testing," 30 *U.C. Davis L. Rev.* 1113, 1115 (1997).
[1004] *Id.*
[1005] See e.g., *supra* Chapter 7: DNA – Probability Estimates Elevated By Smoke and Mirrors to Certainty; Chapter 8: False Positives – DNA Testings Dark Side; and, Chapter 9: A Random Match Probability and False Positive Probability Are Divergent.
[1006] Kurland, *supra* note 23, at 153 (Speech by former FBI Director William Sessions about the certainty of DNA evidence.).
[1007] An example of this is that Timothy Durham's jury chose to believe the testimony of a lone crime lab technician that his DNA matched that of an 11-year-old girl's rapist, over the testimony of 11 alibi witnesses who placed Durham in *another state* at the time of the attack. Subsequent DNA tests proved the experts testimony was inaccurate, which incidentally corroborated the testimony of the 11 alibi witnesses, and Durham was released from prison after four years of wrongful imprisonment. See, Thompson, *supra* note 10, at 48.
[1008] See, *supra* Chapter 4: Doctored Tests and Testimony Undermine the Presumption of Innocence.
[1009] Mary A. Fischer, "The FBI's Junk Science," *Gentlemen's Quarterly*, Jan. 2001, at 115-6.
[1010] See e.g., Underwood, *supra* note 4, at 162..("Lawyers, judges, and critics of expert evidence assume that the public in general, and jurors in particular, accord an "honorific" status to the expert. [FN65]").
[1011] The effect of expert fingerprint evidence on inducing a defendant's to plead guilty was written about as early as 1924. See, Cole, *supra* note 1, at 205.
[1012] *Id.*
[1013] Quoted in, *Id.* at 205.
[1014] See e.g., Underwood, *supra* note 4, at 175-176. ("Perhaps the most distressing aspect of the honesty problem is the complicity of lawyers, particularly prosecutors. If we cannot expect decency from our prosecutors, what can we expect to get from the other side of the "v."? Yet the literature is full of cases in which prosecutors went along with, and in a few cases even solicited, bogus evidence while at the same time withholding exculpatory evidence.") *Id.*
[1015] *Id.* at 153, note 15 ("For evidence that the "junk science" problem is real in the United States and in Commonwealth countries, see David Bernstein, "Junk Science in the United

States and the Commonwealth," 21 *Yale J. Int'l L.* 123 (1996); see also, Paul C. Giannelli, "'Junk Science": The Criminal Cases," 84 *J. Crim. L. & Criminology* 105 (1993). Compounding the problem of prosecution biased testimony is that the reliableness of the tests underlying that testimony can be just as flawed as in the United States. For example, It was reported in Australia in May 2004 that there were inconsistencies in the laboratory practices of Victoria Police lab technicians in testing "DNA, bloodstain and amphetamine" evidence. The problems contributed to incorrect test results for many hundreds of crime evidence samples. "Forensic Work Questioned," *Herald Sun* (Melbourne, AUS), May 18, 2004, referenced at http://www.dnaresource.com/05-212004%20Summary.pdf (last visited September 30, 2004).

[1016] Bruce A. Mcfarlane, "Convicting The Innocent – A Triple Failure of the System," *Canadian Criminal Law*, August 2003, at 57. Available at, http://www.canadiancriminallaw.com/articles/articles%20pdf/Convicting%20the%20Innocent. pdf (last visited Sept. 19, 2004). See also, Underwood, *supra* note 4, at 153, FN. 15 ("For evidence that the "junk science" problem is real in the United States and in Commonwealth countries, see, Bernstein, *supra* note 1010; see also, Giannelli, *supra* note 1010.

[1017] McFarlane, *supra* note 1016, at 57.

[1018] *Id.* Insofar as this concern relates to DNA, considered the gold standard of physical evidence, Cyril H. Wecht, past president of the American Academy of Forensic Sciences wrote in June 2003, "There can be little doubt in the minds of trained, experienced forensic scientists that testing defects, backlog pressures, inadequately qualified personnel, and prosecutorial bias exist in many other DNA labs even though they have not yet been uncovered and publicly reported." Wecht, *supra* note 44, at 1.

[1019] *Id.*

[1020] Yant, *supra* note 20, at 66.

[1021] *Id. at 66..*

[1022] See e.g., Thompson, *supra* note 10, at esp. 53. See also, Kelly, *supra* note 25, at 29-31.

[1023] Underwood, *supra* note 4, at 163.

[1024] See e.g., Thompson, *supra* note 10, at 47, 53. The testing of forensic technicians is neither external (i.e., administered and evaluated by an organization unassociated with the laboratory whose technicians are tested), nor is it blind, nor is it representative of the real-life situations requiring analysis in a typical criminal case. For a contrast to clinical labs, see, Kelly, *supra* note 25, at esp. 29-31. (The performance of technicians in clinical laboratories are evaluated by blind proficiency tests.)

[1025] *People v. Marshall*, No. BA-069796 (Sup.Ct. LA County), Motion To Exclude DNA Evidence, October 10, 1995, at 20.

[1026] National, *supra* note 801, at 89. (Blind proficiency tests need to mimic real-life conditions in every particular, such as a poor quality blood sample, and not one obtained under pristine laboratory conditions.)

[1027] *Id.* at 89. (Such tests need to be representative of real-life cases in regards to sample quality, accompanying descriptiveness, etc.)

[1028] *Id.* at 89.

[1029] *People v. Marshall*, No. BA-069796 (Sup Ct. LA County), Motion To Exclude DNA Evidence, October 10, 1995, at 21 (emphasis added). See also, Teichroeb, supra note 756, at B5 ("The scary part is that the people sitting on juries think all these precautions have been made and they can rely on the evidence," said Roger Hunko, past president of the 750-member Washington Association of Criminal Defense Lawyers.") *Id.*

[1030] See e.g., Thompson, *supra* note 10, at esp. 53.

[1031] National, supra note 801, at 89.

[1032] Thompson, *supra* note 10, at 53.

[1033] For example, in 1999 $147 billion was spent on state and federal law enforcement in the

U.S. The maximum of 3 million that blind proficiency is estimated to cost is 1/149,000th of that amount. 1999 is the latest year that statistics are available from the Bureau of Justice Statistics. For expenditure data see, Sidra Lea Gifford, "Justice Expenditure and Employment in the United States, 1999," *Bureau of Justice Statistics*, NCJ 191746, Feb. 2002, available at, https://www.bjs.gov/index.cfm?ty=pbdetail&iid=1019 (last visited Sept. 20, 2004).

[1034] For an excellent summary of the FBI's decades long opposition to Congressionally mandated external blind proficiency testing that began in at least 1967, and continues to this day, see, Kelly, *supra* note 25, at 29-31. In 1981 the FBI did agree to conduct *internal* proficiency testing. Which as a case of the fox guarding the hen house, could have been predicted to have had no real value. Revelations over more than the past two decades have confirmed that internally monitored technician testing is of no practical value at assuring quality control. See *Id*. at 30-31.

[1035] Fischer, *supra* note 1009, at 116.

[1036] *Id*. at 115-116.

[1037] *Id*. at 115. ("Whitehurst called it "junk science," and most jurors, even most judges, didn't have the knowledge to realize it wasn't real science and were impressed by the agents' authoritative sounding language and by their affiliation with the FBI.")

[1038] See, *supra* Chapter 3.II: Fred Zain – West Virginia's Crime Lab and Bexar County, TX; Chapter 3.IV: Joyce Gilchrist – Oklahoma City's Police Lab, and Chapter 3.X – Montana State Police Crime Lab and Chapter 3.XI – Washington State Patrol Crime Lab.

[1039] Fischer, *supra* note 1009, at 116. ("…most of them [FBI forensic technicians] are not scientists. They are basically people the bureau gets off the street, trains them for a year and then calls them bomb experts.") *Id*.

[1040] Kelly, *supra* note 25, at 29-31, esp. 29.

[1041] *Id*. at 29-31, esp. 29.

[1042] Eric Lander, "DNA Fingerprinting on Trial," *Nature,* 1989 Jun 15;339(6225):501-5, 505. Lander is a mathematician and geneticist at Harvard University and the Massachusetts Institute of Technology's Whitehead Institute, cited in Kelly, *supra* note 25, at 29.

[1043] John Solomon, "Probe of FBI's DNA Lab Practices Widens," *FoxNews.com* (Associated Press), April 29, 2003., Available at, http://www.foxnews.com/story/2003/04/28/probe-fbi-dna-lab-practices-widen.html (last visited September 19, 2004).

[1044] *Id*.

[1045] *Id*.

[1046] This is demonstrated by the fact that proficiency testing, even if it were associated with accreditation (which it typically is not), does not necessarily contribute to ensuring the *accuracy* of test results. See e.g., Thompson, *supra* note 10, at 48. ("…this work [proficiency testing] is designed more to test the uniformity of DNA test results among laboratories using the same protocol than to determine the rate of errors.") *Id*.

[1047] See e.g., the previous sub-section, I – The FBI's Unscientific Crime Lab.

[1048] Fischer, *supra* note 1009, at 149. (In regards to the FBI's accreditation by the ASCLD in 1999, "Janine Arvizu, a nationally known forensic scientist who specializes in auditing labs" called it "a perfunctory exercise. … the ASCLD, whose members are all affiliated with the prosecution, has the least rigorous accreditation program I've ever seen.") quotes at 149. It is worth noting that Jacqueline Blake faked at least 103 DNA tests after the FBI crime lab's accreditation by the ASCLD. "FBI Lab Investigation Widens," *TalkLeft*, April 28, 2003, available at, http://talkleft.com/new_archives/002530.html (last visited September 30, 2004).

[1049] The scientific method does not involve a "single, fixed procedure." Ralph L. Rosnow and Robert Rosenthal, *Beginning Behavioral Research: A Conceptual Primer* (New York: Prentice Hall 2001), 6. Whatever its form in a particular situation, the scientific method's overriding characteristic is being based on *empirical reasoning*, which is identified as deductions based on "logic and the use of controlled observation and measurement. *Id.* at 6.

[1050] *Daubert*, 509 U.S. at 593 (citation omitted).

[1051] See e.g., "Scientific Methods," *supra* note 24, at 119-120. (Among the scientific methods characteristics are: "testable consequences;" "repetition of the test;" and "reliability and accuracy."). This is similar to, but not quite the same as falsification. This was recognized in, *Daubert*, 509 U.S. at 593 (citation omitted). (Testing a methodology for its falsity "is what distinguishes" a scientific analysis from a non-scientific opinion. For example, the basic equation 2+2, explains the principle of this concept. Whether a person is in Tibet, Australia or France, whether in 1805 or 2025, the answer is 4.).

[1052] See e.g., "Scientific Methods," *supra* note 24, at 119-120. (Among the scientific methods characteristics are: "testable consequences;" "repetition of the test;" and "reliability and accuracy."). This was recognized in, *Daubert*, 509 U.S. at 593 (citation omitted). (Testing a methodology for its falsity "is what distinguishes" a scientific analysis from a non-scientific opinion. For example, the basic equation 2+2, explains the principle of this concept. Whether a person is in Tibet, Australia or France, whether in 1805 or 2025, the answer is 4.)

[1053] "Scientific Methods," *supra* note 24, at 119-120. (Among the scientific methods characteristics are: "testable consequences;" "repetition of the test;" and "reliability and accuracy."). See also, *Daubert*, 509 U.S. at 593.

[1054] "Scientific Methods," *supra* note 24, at 119-120. (Among the scientific methods characteristics are: "testable consequences;" "repetition of the test;" and "reliability and accuracy."). See also, Edward J. Imwinkelried, "Evidence Law Visits Jurassic Park: The Far-Reaching Implication of the Daubert Court's Recognition of the Uncertainty of the Scientific Enterprise," 81 *Iowa L. Rev.* 55, 62 (1995) (quotations and citations omitted).

[1055] A court order is necessary for evidence to be made available to a defendant so it can be tested by an outside independent forensic laboratory or examined by a freelance expert. See, Kelly, *supra* note 25, at 15. To ensure accuracy such independent testing would need to entail double blind procedures. *Infra* notes 1191-1212, and accompanying text. (explaining why double blind testing produces more accurate test results than zero or single blind testing procedures).

[1056] Credit must go to Fred Woodworth for applying to fingerprint analysis, what was learned about the horse in the early 1900s known as Clever Hans. See e.g., Woodworth, *supra* note 6, at 49. An excellent account describing the process whereby Clever Hans, a horse with the seeming power to correctly perform complex mathematical computations, was shown to not possess any unique cognitive powers, is, Pfungst, *supra* note 333.

[1057] Pfungst, *supra* note 333, at 20-22, quote at 25..

[1058] *Id.* at 20-21.

[1059] *Id.* at 22.

[1060] *Id.* at 19-22.

[1061] *Id.* at 19.

[1062] *Id.* at 24, 35.

[1063] *Id.* at 141 (Referring to a person asking Hans a question as the "experimenter or questioner.")

[1064] *Id.* at 1, 3. Clever Hans was not unique in that there are many accounts since at least the sixth century of horses able to spell and solve arithmetic problems, and of dogs that could do things such as beat humans at dominoes. *Id.* at 231-233, and also, 177-178. What set Hans apart was mass communication techniques enabled people around the world to learn of his exploits, and he relied on the subconscious movements of a questioner that were so subtle that even those investigating him (The Hans Commission) had difficulty uncovering the secret of his prowess. See e.g., Robert Rosenthal and Lenore Jacobson, *Pygmalion In The Classroom* (Holt, Rinehart & Winston 1968), at 37. Preceding Hans by some years was Kepler, an

Menace To The Innocent

English bull-dog who correctly barked the answer to complex math problems "such as extracting square roots." Pfungst, *supra* note 333, at 177-178, quote at 178. Similar to Hans, his owner, Englishman Sir William Huggins, "was convinced that the dog could see from the questioner's face, when he must cease barking, for he would never for an instant divert his gaze during the process." *Id.* at 178.

[1065] *Id.* at 4.

[1066] *Id.* at 141. ("As a matter of fact, it made no difference who desired an answer, for the only person upon whom the experiment depended was the questioner, that is, the one who asked the horse to tap. We have everywhere designated this person as the experimenter or questioner. It was he who gave the directions, and since all that were involved were visual signs, the drama in which Hans appeared as the hero, was nothing but a pantomime.") *Id.*

[1067] *Id.* at 88. ("Beyond a doubt these necessary signs were given involuntarily by all the person involved and without any knowledge on their part that they were giving any such signals."); see also, Rosenthal, *supra* note 1064, at 36.

[1068] *Id.* at 142. ("Hans, however, was also a faithful mirror of all the errors of the questioner.")

[1069] *Id.* at 61-62. (The highest accuracy was when the questioner stood less than 3-1/2 meters (about 11 feet) from Hans. If the questioner stood 3-1/2 to 4 meters away from Hans, his accuracy fell off significantly to about 40%, and if the questioner was 4 to 4-1/2 meters from Hans, about 33% of his responses were correct.)

[1070] *Id.* at 62. ("When a position immediately behind the horse was taken – a somewhat dangerous proceeding, since Hans would at once begin to kick – no response could be obtained until he succeeded in turning far enough around to get the questioner within view.")

[1071] *Id.* at 62. ("One might even turn his back upon Hans during the tests, for the signal for stopping was not obtained from the face of the questioner, but from a movement of the head.")

[1072] *Id.* at 61-62, esp. at 141 ("All speech was superfluous and, except in so far as the tone of voice in which it was spoken was soothing or reprimanding, it was quite unintelligible to the horse.") *Id.*

[1073] *Id.* at 141.

[1074] *Id.* at 61-62, 141.

[1075] *Id.* at 61-62. (Explains the decrease in Hans' accuracy as the questioner was further away. Also, Hans' accuracy decreased as lighting conditions worsened, such as at dusk). Clever Hans was a fraud in the sense that people ascribed near mystical powers of cognition to him that he didn't possess. *Id.* at 88; see also, Rosenthal, *supra* note 1064, at 36. However there is no reason to believe that Hans or his owner was aware of the deception. *Supra* notes 1066-1068 and accompanying text.

[1076] See e.g., *supra* notes 991-1001 and accompanying text. (The text in the document referring to those footnotes relate how crime lab technicians produce results consistent with the prosecution's theory of a crime.)

[1077] See e.g., *supra* notes 656-672 and accompanying text. (The text in the document referring to those footnotes relate to how the FBI conducted a quasi-test of crime lab fingerprint examiners nationwide in order to establish a zero rate of error, as a defense to a *Daubert* challenge to the scientific soundness of fingerprint evidence in *U.S. v. Mitchell*, CR No. 96-407 (E.D. PA).).

[1078] See e.g., Woodworth, *supra* note 6, at 51. See also, *supra* notes 478-483 and accompanying text. (These can be summarized in the idea that the smaller the area covered by a fingerprint sample, the greater the likelihood a match will be made when it is compared to the complete print samples of other people.)

[1079] *U.S. v Llera Plaza*, No. 98-362-10 (E.D.Pa. 01/07/2002); 2002.EPA.0000003 ¶ 165 <http://www.versuslaw.com>.

[1080] *Id.* at ¶ 165 ("Nine of the thirty-four responding agencies did not make an identification in the first instance.").

[1081] See *supra* Chapter 6.VI - The *Mitchell* Case (1999), and accompanying text referring to the FBI's test of crime lab fingerprint examiners nationwide in order to establish a zero rate of error as a defense to a *Daubert* challenge to the scientific basis of fingerprint evidence in *U.S. v. Mitchell*, CR No. 96-407 (E.D. PA).).

[1082] *U.S. v Llera Plaza*, No. 98-362-10 (E.D.Pa. 01/07/2002); 2002.EPA.0000003 ¶ 268 note 23 <http://www.versuslaw.com>.

[1083] *Id.*

[1084] *Id.* at ¶ 165, 268 note 23.

[1085] This scenario is realistic considering the 26.5% rate of non-unanimity after the initial examinations.

[1086] See *supra* note 64 and accompanying text. (16% average false positive error rate for ASCLD proficiency tests from 1983 to 1991), and note 71 and accompanying text. (17% average false positive error rate for CTS proficiency tests from 1995 to 2001).

[1087] See e.g., Woodworth, *supra* note 6, at 50.

[1088] *Id.* at 50.

[1089] *Id.*

[1090] See, Jenkins, *supra* note 51, at 103. See also, Kelly, *supra* note 25, at 29-30. Although those proficiency tests of crime lab technician skills was conducted over the four year period of 1974 to 1977, the prevalence of erroneous crime lab testing continues. *Supra* notes 57-89 and accompanying text.. See esp., supra note 64 and accompanying text. (16% average false positive error rate for ASCLD proficiency tests from 1983 to 1991), and supra note 71 and accompanying text. (17% average false positive error rate for CTS proficiency tests from 1995 to 2001).

[1091] *U.S. v Llera Plaza*, No. 98-362-10 (E.D.Pa. 01/07/2002); 2002.EPA.0000003 ¶ 165 <http://www.versuslaw.com>. ("Nine of the thirty-four responding agencies did not make an identification in the first instance.").

[1092] For precautions that police are instructed to take in displaying a suspect to a witness in either a photo array or a lineup, see e.g., "Pretrial Identification Procedures Lineups, Showups, Photographic Arrays," *Clark County Prosecutors, For Police Officers*, December 2000 Bulletin, at, http://www.clarkprosecutor.org/html/police/dec00.htm (last visited September 17, 2004).

[1093] *Id.*

[1094] Fischer *supra* note 1009, at 114.

[1095] "Danny McNamee," *supra* note 731. See also, Cole, *supra* note 1, at 282. (Case of Danny McNamee's wrongful conviction).

[1096] Sherrer, *supra* note 665, at 10.

[1097] *Id.* at 10.

[1098] Jenkins, *supra* note 51, at 103. (Citing proficiency test error rates for crime lab technicians.)

[1099] There are numerous explanations of the Prosecutor's Fallacy as it relates to different types of evidence. See e.g., Thompson, *supra* note 10, at 52. For an excellent lay person's explanation of the prosecutor's fallacy with examples, see "Prosecutor's Fallacy," *Wikipedia.org*, at, http://en.wikipedia.org/wiki/Prosecutor's+fallacy (last visited September 19, 2004).

[1100] *Id.*

[1101] *Id.* The concept of a false positive analysis also applies to other forms of evidence, such as erroneous eyewitness testimony. See, e.g., Donald S. Connery (ed.), *Convicting the Innocent: The Story of a Murder, a False Confession, and the Struggle to Free a "Wrong Man"* (Brookline Books 1996), at 115. ("I read somewhere that Elizabeth Loftus, who is one of the preeminent authorities on eyewitness identification, said that the miracle is when you get it right, not that you get it wrong." *Id.* at 115).

[1102] See e.g., Lempert *supra* note 806.

[1103] See e.g., *Id.*

[1104] For an indication of the prevalence of a false positive analysis in proficiency tests under conditions significantly less demanding than a "real-world" analysis, see *supra* notes 50-89 and accompanying text. In regards to the prevalence of an witness' false positive analysis of a crimes perpetrator, see e.g., Elizabeth Loftus, *Eyewitness Testimony* (Harvard Univ. Press 1996 ed.) In the Preface the author writes: "A major reason for my writing this book has been a long-standing concern with cases in which an innocent person has been falsely identified, convicted, and even jailed." *Id.* at xi. See also, Connery, *supra* note 1101, "I read somewhere that Elizabeth Loftus, who is one of the preeminent authorities on eyewitness identification, said that the miracle is when you get it right, not that you get it wrong." *Id.* at 115.

[1105] *Id.* See also, *supra* notes 817-825 and accompanying text.. Those notes relate to how random match (coincidental) probability can be used by a prosecutor to obscure that the false positive probability in a case that has seemingly rock solid expert testimony can be expected to be 10%. (The author does not know the "random match probability" that was testified to at Sutton's trial. However a billion to one is being used to demonstrate that even with a high probability that a coincidental match is unlikely, the false positive probability can still be so low as to undercut by a factor of 10, the 99% probability of guilt suggested by the Supreme Court in *Schlup* 513 U.S. at 324-325, is necessary to sustain a finding of guilt.)

[1106] As of November 2003, no state or federal court in the country requires a jury to be told of the false positive probability associated with DNA evidence, even though in a 1992 report the National Research Council stated, "…laboratory error rates must be continually estimated in blind proficiency testing and must be disclosed to juries." National, *supra* note 798, at 89.

[1107] *Id.*, See also, Lempert *supra* note 806.

[1108] *Id*

[1109] See e.g., Thompson, *supra* note 10, at esp. 48. (Discusses Timothy Durham's case.)

[1110] The 10% figure is the posterior odds calculated from analyzing the relationship between the Prior Odds, the Random Match Probability and the Probability of a False Positive. See how these relationships applied to Josiah Sutton case at *supra* notes 818-820 and accompanying text. (See e.g., Thompson, *supra* note 10, at Table 1: Posterior odds that a suspect is the source of a sample that reportedly has a matching DNA profile, as a function of prior odds, random match probability, and false positive probability.) A twist is what can be called the *expert's fallacy*. An example of this occurred in December 1993 when the U.K.'s Criminal Court of Appeal quashed Andrew Deen's rape conviction based on the jury's contamination by the forensic witnesses' testimony that since he calculated the odds were three million to one semen found at the crime scene wasn't Deen's, then the odds he wasn't guilty were likewise three million to one. See also, *The Prosecutor's Fallacy*, The Sixth Form College, at: http://www.colchsfc.ac.uk/maths/dna/discuss.htm (last visited September 19, 2004). (Discusses that in spite of what is known about the prejudicial effect on jurors of exposure to the prosecutor's and/or expert's fallacy, no U.S. court is known to have reversed a conviction for that reason.)

[1111] For an account of Sally Clark's case, see, Hans Sherrer, "Sally Clark's Conviction of Murdering Two of Her Children is Quashed After Discovery the Prosecution Concealed Evidence of Her Innocence," *Forejustice.org*, February 22, 2003, available at, http://forejustice.org/wc/sally_clark_freed.htm. (Last visited September 16, 2004). Wikipedia, a World Wide Web encyclopedia uses Sally Clark's case to illustrate the prosecutor's fallacy. See, "Prosecutor's Fallacy," *supra* note 1099.

[1112] *Id.* The death of Sally Clark's first infant son, Christopher, was originally attributed to a "lower respiratory tract infection." *Id.* It was reclassified as death by smothering when her second infant son's death was classified as due to a "non-accidental injury consistent with shaking." *Id.* Sally Clark did not present the defense that the boy's died from "cot death," but because she did claim they both suddenly stopped breathing, the prosecution attempted to undermine her defense by presenting evidence of the low probability that two children in the same family would die from SIDS. *Id.*

[1113] *Id.* "Cot death" is a British term describing what is known in the U.S. as SIDS – Sudden Infant Death Syndrome.

[1114] "Sally Clark: More About the Statistics and the Appeal," *Sally Clark website,* http://www.sallyclark.org.uk/AppealStats.html. (Last visited December 7, 2003). (Noting the prosecution's repeated references to the one in 73 million probability during the trial and in summing up the case to the jury).

[1115] *Id..*

[1116] Sherrer, *supra* note 1111.

[1117] Professor Hill calculated the probability was as low as 1 in 60 and as high as 1 in 130. The median is 1 in 95. See, Helen Joyce, "Beyond Reasonable Doubt," *+Plus magazine,* September 2002 (Issue 21), at, http://pass.maths.org.uk/issue21/features/clark (Last visited September 9, 2004).

[1118] Professor Hill calculated the probability was as low as 1 in 60 and as high as 1 in 130. *Id.* The median is 1 in 95. See, *Id.* The prosecutor argued to the jury the probability of two children in one family dying of natural causes was 73 million to 1. Thus, 73,000,000/95=768,421.

[1119] Sherrer, *supra* note 1111.

[1120] *Id.*

[1121] *Id.*

[1122] *Id.*

[1123] Sir Roy Meadows, the paediatrician expert who testified in Sally Clark's case that the odds of a second baby in a family dying from "cot death" (SIDS) after a previous death due to the syndrome is one in 73 million, also provided the critical testimony against Angela Cannings – whose conviction of murdering two of her children was quashed in Dec. 2003. Press Association, "Mother Cleared of Infant Sons' Deaths," *The Guardian,* Dec. 10, 2003, at, http://www.guardian.co.uk/uk_news/story/0,3604,1104001,00.html. (last visited Sept. 19, 2004). Demonstrating that public exposure of "junk science" techniques can have an effect on jury verdicts, on June 11, 2003 Trupti Patel was acquitted by a jury in just 90 minutes of murdering three of her children based on similar evidence used to convict Sally Clark and Angela Cannings. See, Staff, "Mother Cleared of Killing Babys," *BBC News,* June 11, 2003, at, http://news.bbc.co.uk/2/hi/uk_news/england/berkshire/2982148.stm (last visited Sept. 19, 2004). See also, James Le Fanu and David Derbyshire, "In the Rush to Protect Children, 'Experts' Use Junk Science To Accuse Innocent Parents," *The Daily Telegraph,* Dec.13, 2003, at, http://www.telegraph.co.uk/news/main.jhtml?xml=/news/2003/12/13/nbaby13.xml (last visited Sept. 19, 2004).

[1124] Since the prosecutor's fallacy is based on erroneous reasoning about the probative value of alleged prosecution evidence and/or testimony, there is nothing preventing it from being used with virtually any type of alleged evidence.

[1125] See e.g., Loftus, *supra* note 1104. (In her Preface the author writes, "A major reason for my writing this book has been a long-standing concern with cases in which an innocent person has been falsely identified,convicted, and even jailed." at xi). See also, Connery, *supra* note 1101, at esp. 115 ("I read somewhere that Elizabeth Loftus, who is one of the preeminent authorities on eyewitness identification, said that the miracle is when you get it right, not that you get it wrong." *Id.* at 115).

[1126] Loftus, *supra* note 1104. ("Eyewitness testimony is among the most damning of all evidence that can be used in a court of law. When an eyewitness points a finger at a defendant and says, "He did it! I saw him. I was so shocked I'll never forget his face!" The case is as good as over. "Cast-iron, brass-bound copper- riveted, and airtight," as one prosecutor put it. The defendant sits helpless, without hope, fear turning into panic. Only someone has been accused of a crime he didn't commit can know just how devastating the experience can be. I once heard a falsely accused person say, "I'd rather have terminal cancer than go through this." *Id.* at v-vi.)

[1127] In Sally Clark's case e.g., the prosecutor's fallacy argued to the jury overstated by a factor of 768, 421, the probability that two children in a family wouldn't die of natural causes. See, Joyce, *supra* note 1112. (Professor Hill calculated the probability was as low as 1 in 60 and as high as 1 in 130. The median is 1 in 95. The prosecutor argued to the jury the probability of two children in one family dying of natural causes was 73 million to 1. Thus, 73,000,000/95=768,421.)

[1128] National, *supra* note 798, at 89.

[1129] *People v. Marshall*, No. BA-069796 (Sup.Ct. LA County), Motion To Exclude DNA Evidence, October 10, 1995, at 16 (emphasis in original).

[1130] Richard Lempert, "Some Caveats Concerning DNA As Criminal Identification Evidence: With Thanks to the Reverend Bayes," 13 *Cardozo L.Rev.* 303, 325 (1991) (the probability of a coincidental match between people who have the same DNA profile "is usually dwarfed by the probability of a false positive error.") (Cited in *People v. Marshall*, No. BA-069796 (Sup.Ct. LA County), Motion To Exclude DNA Evidence, October 10, 1995, at 17.

[1131] Paul J. Hagerman, "DNA Typing in the Forensic Arena," 47(5) *Am. J. Hum. Genet.* 876-877, Nov. 1990.

[1132] Lempert, *supra* note 1130, at 325 (the probability of a coincidental match between people who have the same DNA profile "is usually dwarfed by the probability of a false positive error.") (in binder). See also, R.C. Lewontin & Daniel Hartl, "Population Genetics in Forensic DNA Typing," 254 *Science* 1745, 1749 (1991) (in binder) ("The rate of false positives defines a practical lower bound on the probability of a match, and probability estimates based on population data that are smaller than the false-positive rate should be disregarded.") (Cited in *People v. Marshall*, No. BA-069796 (Sup.Ct. LA County), Motion To Exclude DNA Evidence, October 10, 1995, at 19.

[1133] For an enlightening discussion of the factors surrounding the effectiveness of the prosecutor's fallacy, see, Thompson, *supra* note 10, at 51-52. (Discusses the prosecutor's fallacy and the importance of knowing the posterior odds of a false match in making an assessment of the probability of a coincidental match between a suspect's DNA and crime scene DNA.)

[1134] This is not an idle question. At least one former FBI crime lab technician, Bill Tobin, wonders if some technicians are so unknowledgeable about what they *should* be doing that they may not meet the legal standard of criminal intent. Fischer *supra* note 1009, at 148-9.

[1135] *Federal Rules of Evidence*, Rule 102 – Purpose and Construction. [hereinafter *Fed. R. Evi.* 102] (The stated purpose Federal Rules of Evidence is, "That the truth may be ascertained and the proceedings justly determined.").

[1136] *Id.* (The stated purpose Federal Rules of Evidence is, "That the truth may be ascertained and the proceedings justly determined.").

[1137] *Id.*

[1138] Fred Zain, for example, was prosecuted by the State of West Virginia for perjury in 1994, but the indictment was dismissed due to expiration of the statute of limitations. See, "Death of Lying Chemist Fred Zain," *supra* note 181.

[1139] See e.g., *Ratzlaf v. U.S.*, 510 U.S. 135, 141 (1994)

[1140] A strict liability statute merely requires the commission of an act without regard for the mental intent of the actor.

[1141] See e.g., Sherrer, *supra* note 114, at 258. ("A de facto code of silence contributes to hiding the illegal and amoral actions committed by members of a bureaucracy.")

[1142] See e.g., *Ratzlaf*, 510 U.S. at 141 (Willfulness defined).

[1143] Swearington , *supra* note 114, at 55. (J. Edgar Hoover let it be known he would not tolerate "… any whistleblowers in the FBI."). Members of a bureaucracy are likened to belonging to an identifiable "bureaucratic brotherhood," in Sherrer, *supra* note 114, at 250.

[1144] *Id.* at 55.

[1145] There is no known instance of this actually happening, because no FBI crime lab technician is known to have been federally prosecuted for perjury. Perhaps the closest an agent has come to providing evidence of a technician's perjury was in 1989 when Fredrick Whitehurst provided the defense in the prosecution of Steve Psinakis in San Francisco, with information that the FBI's expert lab witness was offering an investigative opinion that was not scientifically supported, concerning the "explosives-residue evidence." Using that "inside" information, Psinakis' lawyers were able to undermine the expert's testimony and the jury acquitted him of all charges. See, Kelly, *supra* note 25, at 37-43. Seven years later Whitehurst was forced out of the FBI after blowing the whistle to Congress on serious deficiencies in the FBI crime labs operation. See, Johnston, *supra* note 111, and, Cannon, *supra* note 113.

[1146] Fischer *supra* note 1009, at 148.

[1147] Kelly, *supra* note 25, at 20-21.

[1148] *Id.* at 20-21.

[1149] *Id.* at 20-21.

[1150] This is one way that technicians engaging in widespread insubstantial testimony, such as FBI lab employee Tom Curran, and state workers Fred Zain, Joyce Gilchrist and Arnold Melnikoff, were able to operate below the radar screen of public disclosure for many years.

[1151] Fischer *supra* note 1009, at 148.

[1152] Lefcourt, *supra* note 2.

[1153] In every case that this occurs, the defendant is the victim of a prosecutor orchestrated frame-up. See e.g., See also, Sherrer, *supra* note 34; see also, "Prosecutorial Lawlessness," *supra* note 34.

[1154] *Id.*

[1155] Fischer *supra* note 1009, at 148-9. (""In the case of Webb and other agents," says Bill Tobin, "there is a serious question of intent.")

[1156] *Id.* at 148.

[1157] *Id.* at 148-9. (""In the case of Webb and other agents," says Bill Tobin, "there is a serious question of intent.")

[1158] *Id.* at 148-9.

[1159] *Id.* at 148-9.

[1160] *Id.* at 149.

[1161] See e.g., *Id.* at 148.

[1162] *Id.* at 148.

[1163] See e.g., *Id.* at 116. (Quoting former FBI chief metallurgist Bill Tobin, "…you've got a person who has the aura of the FBI surrounding him, and anything he says or does is not questioned.")

[1164] *Id.* at 116.

[1165] In the course of researching this article, the author was unable to find a single federal perjury prosecution of an FBI lab technician for untruthful courtroom testimony. This is supported by the finding of the authors of Kelly, *supra* note 25, at 14. In a rare federal prosecution of an FBI lab technician, but which didn't involve courtroom perjury, on May 18,

2004, Jacqueline Blake pled guilty in federal court in Washington D.C. to one count of making false statements on official government reports, although she acknowledged in her plea agreement she knew that any one of her more than 100 false DNA certifications from August 1999 to June 2002 could have been used to identify a suspect in a criminal investigation and influence trial testimony. Anderson, *supra* note 145.

[1166] The DOJ declined to prosecute the FBI's lab technician after she admitted in writing that she knowingly gave false testimony during a pre-trial hearing in a state murder case. Evans, *supra* note 131.

[1167] Evans, *supra* note 131.

[1168] *Id.* (After the DOJ declined to prosecute, Kentucky state prosecutors charged the technician, Kathleen Lundy, with false swearing. It was reported in April 2003 that she would plead guilty to the misdemeanor charge. *Id.* On June 17, 2003, Lundy pled guilty to false swearing and was given a 90-day suspended sentence and fined $250. Evans, *supra* note 138. The Fayette County prosecutor said after her sentencing that he he had recommended, "She's already lost her job and paid severely, through the loss of her job and her reputation," Smith said. "In my mind, that was sufficient. I didn't see that the taxpayers of Fayette County needed to keep her up for a while." *Id.*

[1169] The procurement of perjury is criminalized in 18 U.S.C. §1622 – Subornation of perjury – "Whoever procures another to commit any perjury is guilty of subornation of perjury ..."

[1170] This is the English translation of the original Latin, *Sed quis custodiet ipsos custodes?* From a satire by Juvenal (Decimus Junius Juvenalis 55-127 AD). See, http://www.barbeleis.de/custodes/

[1171] Since lab personnel commit potential crimes while providing expert support for the prosecution's theory of a case, it is not in the interest of prosecutors to criminally charge them for doing so – since that could potentially undermine the conviction. Falling under the exception to the rule category, there have been a few rare instances when a crime lab technician has been charged with a crime related to provably perjurious testimony. An example of this is the case of FBI lab technician Kathleen Lundy, who admitted in writing that she deliberately gave false testimony during a pretrial hearing for a murder case in Kentucky. After the U.S. Attorney's Office declined to prosecute her, the State of Kentucky charged her with misdemeanor false swearing. See, Evans *supra* note 131, and Evans, *supra* note 138.

[1172] Although there have been some half-hearted proposals that might create the appearance of changing the operation of the FBI's crime lab, they would in fact have no more real-world impact on the lab's operating procedures than the 1999 accreditation of the lab by the ASCLD had. See, Fischer, *supra* note 1009, at 149.

[1173] *Fed. R. Evi.* 102 (The stated purpose Federal Rules of Evidence is, "That the truth may be ascertained and the proceedings justly determined.") *Id.*

[1174] Diogenes was a Greek philosopher (412-323 B.C.), who according to legend wandered the streets of Athens in the daytime carrying a lantern in search of an honest person. *Diogenes*, Encyclopædia Britannica Online, http://search.eb.com/eb/article?tocId=9030530 (last visited on September 19, 2004).

[1175] See e.g., Lyn Haber and Ralph Haber, "Double-Blind Procedures in Forensic Identification and Verification," *Human Factors Consultants*, March 9, 2002, at 1-3, 13, unpublished article available at, http://humanfactorsconsultants.com/research.html (last visited September 16, 2004). (Explaining zero blind, single blind and double blind procedures as they apply to crime laboratory technicians.).

[1176] See e.g., *Id.* at 1-3, 13.

[1177] *C.F., The Oxford English Dictionary, supra* note 460, definition of *Blind, 9c* ("Applied to (the conduct of) a test or experiment in which information about the test that might lead to bias in the results is concealed from the tester or the subject (or both) until after the test is made, esp. as *blind testing,*") *Id.*

[1178] See e.g., Haber, *supra* note 1175, at 1-3, esp. 13 ("The expected outcome is that very high performance will be found."), unpublished article available at, http://humanfactorsconsultants.com/research.html (last visited January 12, 2004).

[1179] See e.g., *Id.* at 3 ("The effect of experimenter bias, as well as subject expectation, has been demonstrated with the same rigor and scientific validity as the earth's rotation around the sun.") *Id.*

[1180] See e.g., Arthur K. Shapiro, M.D. and Elaine Shapiro, M.D., "The Powerful Placebo," *The John Hopkins University Press* (Baltimore 1997), at 154-5, 168-9, 173-4, esp. 154 ("[Dr. Stewart] Wolf noted that every medication given to [a patient named] Tom caused a physostigmine effect, which led him to use placebos in subsequent studies because *the placebo effect could outweigh the ordinary effect of drugs.* This was the first demonstration of the observable, measurable effect of placebos. Other papers provided *documentation attesting that placebos could induce toxic effects.*" emphasis added) *Id.*

[1181] Gary Greenberg, "Is it Prozac? Or Placebo?," *Mother Jones*, Nov./Dec. 2003, p. 76 (6), esp. 78. (It was only due to FDA mandated double-blind tests that the authors were able to discover the ineffectiveness of anti-depressants compared to the expectations of people who take an inert placebo. See also, Ed Cohen, "The Placebo Disavowed: Or Unveiling the Bio-Medical Imagination," *The Yale Journal of Humanities in Medicine*, at 6, http://info.med.yale.edu/intmed/hummed/yjhm/archives/ecohenprint.htm (last visited April 11, 2004). (It has been observed that, "within scientific medicine, the "placebo effect" is bracketed not because it does not produce healing effects, but precisely *because it does.*" *Id.* at 6. That same article explains the first known recognition of the "placebo effect" was in the 1784 report of a committee commissioned by France's King Louis XVI to study the unorthodox, but effective healing techniques of Austrian Physician Franz Anton Mesmer: "the *imagination* is the true cause of the effects attributed to magnetism." *Id.* at 9, (citation omitted) (emphasis added). The first recorded use of blind testing was by that same Royal Commission, "the commissioners sought to eliminate any possibility that the subjects of their examination could be influenced by factors other than the single agent whose efficacy they were charged to evaluate.") *Id.* at 8. In 1785, a year after the report was issued, the word "placebo" was first used to describe "materialization of a non-determinant cause." *Id.* at 8.)

[1182] *The Oxford English Dictionary, supra* note 460, defines "*self-fulfilling prophecy*: a prophecy or prediction which gives rise to actions that bring about its fulfillment."

[1183] Haber, *supra* note 1175, at 11. (In the context of a zero-blind line-up proceeding, "Research evidence has shown that zero blind procedures ... results in many erroneous identifications – as high as 100%, even when the perpetrator is not present in the lineup.") *Id.*

[1184] An example of this is that in the FBI's *Mitchell* (1999) case test of crime lab fingerprint proficiency, 73.5% of the labs initially "matched" the sample identified as the suspect's with the alleged crime scene fingerprint. Doubt that the prints actually matched is indicated by the 26.5% of labs that did not initially succumb to the FBI's overt suggestions, by declaring they didn't match. *Supra* notes 633-649 and accompanying text. That test result was consistent with the what has been known for decades, zero-blind procedures are inherently unreliable, see e.g., Haber, *supra* note 1175, at 11. (In the context of a zero-blind line-up proceeding, "Research evidence has shown that zero blind procedures ... results in many erroneous identifications – as high as 100%, even when the perpetrator is not present in the lineup.") *Id.*

[1185] *Id.* at 11. (In the context of a zero-blind line-up proceeding, "Research evidence has shown that zero blind procedures ... results in many erroneous identifications – as high as 100%, even when the perpetrator is not present in the lineup."). *Id.* The ability of a conscientious crime lab technician and other prosecution experts to convince him or herself of the correctness of a scientific test and/or their testimony, is similar to the placebo phenomena that occurs when a person is convinced that the effect they experience from taking an inert substance was caused by the drug that they *only thought* they were taking. See, Shapiro, *supra* note 1180, at 142-151. It can almost be likened to a form of self-hypnosis.

[1186] *Supra* notes 60-84 and accompanying text.

[1187] Haber, *supra* note 60, at 12.

[1188] *Id.* at 12-13 (emphasis in original).

[1189] *Id.* at 12.

[1190] *Id.* at 12 ("... single blind procedures uncover significantly more errors than zero blind.").

[1191] *Id.* at 12 (Not only do crime labs not use single-blind techniques for arriving at initial test results, but they don't use them to verify test results: "crime laboratories that do verify identifications follow zero blind procedures exclusively.").

[1192] See e.g., Haber, *supra* note 1175, at 13. For an explanation of single-blind testing in the context it is typically associated with, medical testing, see, Shapiro, *supra* note 1180, at 137 ("The earliest technique used was the single-blind procedure. In this paradigm the physician, or investigator knows that control substances are being used and knows which patients are receiving them, but the patient does not know.") *Id.*

[1193] See e.g., *Id.* at 13.

[1194] *Supra* note 1193 and accompanying text, and *infra* note 1195 and accompanying text.

[1195] For the same reason that physicians need to be blinded to which test sample in a study is real and which is a placebo, in order to prevent them from biasing a test's outcome, lab supervisors, police investigators and prosecutors need to be kept in the dark about what sample tested by a crime lab is the actual evidence in a case. See e.g., Shapiro, *supra* note 1180, at 142 (In regards to a medical study that began in 1932 it is written, "As the study progressed ... they [the researchers] realized that the physicians administering the drug were asking the patients "leading questions" about its effect on the severity of pain and thus could be biasing the questions" about its effect on the severity of pain and thus could be biasing the results. A review of the charts revealed that patients were giving contradictory answers to the physicians' questions, making it essential to blind the physicians. The authors emphasized the need for objectivity: "The method were employed for determining cause and effect was more objective and relatively free of personal judgments.") *Id.* See also e.g., *supra* notes 1051-1070 and accompanying text. (Explaining how in the early 20th century subconscious cues were transmitted to the horse Clever Hans by the person asking him a question, who would almost unfailingly lean forward almost imperceptibly when the horse was correctly answering, and lean back when he had tapped out the correct answer.).

[1196] See e.g., Shapiro, *supra* note 1180, at 142-155, esp. 154.

[1197] See e.g., Haber, *supra* note 1175, at 3, 12-13, esp. 12 ("... double blind test procedures uncover significantly more errors than single blind...") *Id.*

[1198] See e.g., *Id.* at 3, 12-13 (Discusses double-blind testing in the context of medical trials for which it is an essential part of the process of ensuring reliable and unbised test results.) See also, Shapiro, *supra* note 1180, at 142-151, esp. 169 (A researcher criticized the use of double-blind techniques precisely because of the results they achieved, after discovering, "that if the double-blind method was used to evaluate an effective accepted drug, an unknown drug, or a placebo, it was impossible to distinguish in terms of effectiveness or action the three types of medication under investigation.") *Id.*

[1199] (Emphasis added to original)

[1200] See e.g., Haber, *supra* note 1175, at 13. (In the context of a proficiency test, "the laboratory's supervisors do not know the correct answers"); See also, Shapiro, *supra* note 1180, at 142 (In the context of a medical drug trial, "a review of the charts revealed that patients were giving contradictory answers to the physicians' questions, making it essential to blind the physicians. The authors emphasized the need for objectivity. ... In summary ... "The data was obtained in a manner relatively free of bias by the use of the [double-]"blind test."") *Id.*

[1201] See also, Shapiro, *supra* note 1180, at 143 (In regards to a medical trial conducted in the 1930s, "Patients detected differences between the placebo and the active drug and wanted the active drug. The investigators recognized that this introduced a subjective bias on the patient's part, and that it was, therefore, *essential to use identical agents.*" emphasis added)

[1202] See e.g., Cohen, *supra* note 1181, at 6.

[1203] *Id.* at 6. (Quoting Isabelle Stengers, *The Invention of Modern Science* (Minneapolis: Univ. of Minnesota Press, 2000), "...the "power of fiction" is constitutive of science generally since "it is that *against which* science must differentiate itself, and that *through which* it defines – disqualifies everything that is not science." (emphasis in original)) *Id.* at 6. Unlike double-blind procedures, zero and single blind testing techniques do not, and cannot protect their results from being influenced by the "power of fiction." *Id.* at 6. Thus an argument can be made that they are not scientific processes – but merely masquerade as such.) *Id.*

[1204] See, *supra* notes 1175-1195 and accompanying text, and also *infra* notes 1208-1209 and accompanying text. (Crime lab testing techniques are a form of what Nobel Prize winning Chemist Irving Langmuir described as "pathological science," because they generate unreliable results that are pawned off as substantial, whether they are or not. That is, they are a form of "the science of things that aren't so.") *Id.*

[1205] Haber, *supra* note 60, at 11. See also, Shapiro, *supra* note 1180, at 173 ("...by about 1980, use of the double blind in studies was essentially required to satisfy the criterion of minimizing bias. Applications for FDA approval of new drugs, with rare exceptions, now require studies using the double-blind method as well as randomization and other methodological safeguards.") *Id.*

[1206] See e.g., Press Release ("UCI Professor"), *supra* note 374. (Relates that a Houston PD Crime lab technician erred in testifying that Josiah Sutton's DNA "definitely" matched that of a car-jacking rapist.) *Id.*

[1207] Shapiro, *supra* note 1180, at 142, 152, 154, esp. 152. (The first use of the phrase "blind test" in the context of a blind or controlled scientific experiment is believed to have been in 1917 by Pharmacology Professor Torald Sollman. However use of the phrase stuck in 1937.) *Id.*

[1208] *Id.* at 154.

[1209] *Id.* at 154.

[1210] *Id.* at 154.

[1211] See e.g., Bedau, *supra* note 25, at 147-148. (Lloyd Miller was wrongly convicted and sentenced to death in 1956 on the basis of insubstantial expert technician testimony that Miller's blood was on a piece of clothing allegedly linked to the murder of a young girl, when it was actually red paint. An independent laboratory later proved the substance was paint, and the clothing was likewise established to not be his.) *Id.*

[1212] See *supra* notes 1156-1175, esp.1174-1175 and accompanying text. (Relating how the horse known as Clever Hans was able in early 20th century Germany, to correctly answer complex mathematics and language questions by reading the answers subconscious "telegraphed" by the person asking him the questions.)

[1213] Haber, *supra* note 60, at 12-13 (Not only are zero-blind procedures used in the initial crime lab testing process, but if any verification is performed, it also follows a zero-blind protocol.).

[1214] Cohen, *supra* note 1181, at 6. ("...the "power of fiction" is constitutive of science generally since "it is that *against which* science must differentiate itself, and that *through which* it defines – disqualifies everything that is not science." (emphasis in original)) *Id.*

[1215] *Supra* notes 1175-1180 and accompanying text.

[1216] *Supra* notes 60-80 and accompanying text. (Relating the ASCLD fingerprint proficiency test results from 1983 to 1991, and the CTS test results from 1995 to 2001).

[1217] See e.g., *infra* notes 1254-1257 and accompanying text. ("Cowboy" is used in the context of its derogatory meaning of "a reckless or irresponsible person." *Random House Webster's Unabridged Dictionary, supra* note 470, *Cowboy:*4. informal).

[1218] Each person this testimony is used against is legally presumed to be innocent at the time it is given in court, and as related herein, numerous cases demonstrating that all too often the person is actually innocent.

[1219] Baden, *supra* note 46, at 232.

[1220] *Id.* at 232.

[1221] *Supra* notes 807-809 and 1002-1004 and accompanying text.

[1222] *Supra* notes 1181-1197 and accompanying text.

[1223] Baden, *supra* note 46, at 232.

[1224] For illustrative examples see the discussion of Fred Zain, *supra* notes 162-184 and accompanying text, and Joyce Gilchrist, *supra* notes 198-215 and accompanying text.

[1225] Baden, *supra* note 46, at 232.

[1226] Robert L. Park, *Voodoo Science: The Road from Foolishness to Fraud*, (New York: Oxford University Press, 2000), at 212. ("Most people who are drawn to voodoo science simply long for a world in which things are some other way than the way they are. Some cannot accept that we are prisoners of the Sun. They look wistfully at the stars that fill the night and imagine that there *must* be some way to overcome the limitations of space and time.") *Id.* at 212.

[1227] Baden, *supra* note 46, at 232. ("Much as I hate to admit it, the sad fact is that some forensic scientists do, indeed, fool a lot of the people a lot of the time.") *Id.*

[1228] A defendant's frame-up by physical evidence that has no actual probative value can provide a vital piece of the puzzle constructed by a prosecutor to ensure the conviction of a defendant - who may or may not be innocent. For an explanation of the prosecutor's role in this process see, Sherrer, *supra* note 34.

[1229] This phenomena was recognized nearly 100 years ago in *The Red Thumb Mark*, R. Austin Freeman's 1907 book about how Scotland Yard's incorrect analysis of a thumb-print incorrectly implicated an innocent man in a theft. Freeman, *supra* note 542. Also, expert testimony for the prosecution is normally so persuasive that it seals a conviction. In 1924 it was observed in *Fingerprint and Identification Magazine*, "we are impressed by the large proportion of cases in which criminals confess when they learn that the finger-print system is being used. ... In one large Midwestern city, criminal lawyers refuse to take cases in which finger-print evidence figures. They cannot afford to risk their reputations on cases which will surely find their clients guilty." Cole, *supra* note 1, at 205.

[1230] Park, *supra* note 1221, at 212.

[1231] See *supra* notes 50-89 and accompanying text. (This is indicated by the high rate of errors by state and federal crime lab tests and technicians in proficiency tests during the last four decades.)

[1232] Baden, *supra* note 46, at 232.

[1233] *Supra* notes 1175-1189 and accompanying text. (Describing zero and single-blind test procedures and their inherent deficiencies.)

[1234] Baden, *supra* note 46 at 232 (Observing that we must not fool ourselves and we are the easiest to fool.)

[1235] *Id.* at 232. The authors note, "Junk science is not new." *Id.* at 233. For a article explaining how 'junk science' is used to prosecute innocent parents for the death of a child, see, *In the rush to protect children, 'experts' use junk science to accuse innocent parents, supra* note 1095. Although some crime lab test procedures may be methodologically sound, but the accuracy of any given result is unknown by the failure to use a reliable double-blind testing protocol.

[1236] See e.g., Kelly, *supra* note 25, at 26, ("The inability of courts to tell the difference between real and junk science was partially responsible for what seems like downright laxity when faced with the shortcomings of forensic examiners.").

[1237] *Id.* at 26.

[1238] *Daubert*, supra at 589. The rule referred to is *Fed. R. Evi.* 702.

[1239] *U.S. v Llera Plaza*, No. 98-362-10 (E.D.Pa. 01/07/2002); 2002.EPA.0000003 ¶¶ 114-123 <http://www.versuslaw.com>.

[1240] *Id.* at ¶¶ 114-123 (Citing *U.S. v. Mitchell*, CR No. 96-407 (E.D. PA)). Test. Budowle, Tr. July 9, 1999, at 122-123, quoted in Gov't Mot. & Resp. at 42-43.)

[1241] *Id.* ¶ 114.

[1242] *Id.* ¶¶ 120-121. (Citing *U.S. v. Mitchell*, CR No. 96-407 (E.D. PA)). Test. Budowle, Tr. July 9, 1999, at 122-123, quoted in Gov't Mot. & Resp. at 42-43 (emphasis added).)

[1243] The word processing program is the largest selling such program in the world, and it is produced by the the largest software company in the world. Both the company and the product will remain unnamed.

[1244] Up to the writing of this article the word processing program's spell-check process was believed by the author to have worked as expected at detecting a word in a document that was not in the spell-check database, either due to being misspelled or a new word that was correctly spelled.

[1245] That competing program will also remain unnamed, but it is supported by a major software company, and as of October 2004, is available on the WorldWideWeb for free downloading.

[1246] *U.S. v Llera Plaza*, No. 98-362-10 (E.D.Pa. 01/07/2002); 2002.EPA.0000003 ¶¶ 120-121 <http://www.versuslaw.com> (Citing *U.S. v. Mitchell*, CR No. 96-407 (E.D. PA)). Test. Budowle, Tr. July 9, 1999, at 122-123, quoted in Gov't Mot. & Resp. at 42-43 (emphasis added).)

[1247] The spell check of the document by a competing and unrelated independent word processing program was the functional equivalent of verifying the spelling in the document by an independent double-blind test. Spelling checking the document by the same word processing program installed on different computers accomplished nothing other than to confirm the erroneous report that there were no spelling errors. That program's masking of the actual spelling errors was the equivalent of a zero-blind test that *never* would have detected the spelling errors, since the program's spell-check protocol simply would have endlessly confirmed, quite incorrectly, that no errors existed. For an excellent analysis of how double-blind testing can provide the same error detection for forensic laboratory tests, see, Haber, *supra* note 1147.

[1248] It is not being intimated in any way that Dr. Dudowle's perjured himself during his testimony. However cast in the most favorable light that testimony was naïve and uninformed. Although its basis was suspect, that did not prevent the testimony from having a possibly prejudicial impact on the judge's opinion in *U.S. v. Mitchell*, since his ruling supporting the admissibility of expert fingerprint testimony was consistent with Dr. Dudowle's attitude that fingerprint analysis methodology is infallible. *U.S. v. Mitchell*, CR No. 96-407 (E.D. PA).

[1249] *Supra* notes 1242-1247 and accompanying text. (Describing that a computer spell checking program can systematically return erroneous results, that are only detected and corrected by a completely separate program.).

[1250] *Supra* notes 1237-1242 and accompanying text. (Describing that a computer spell checking program can systematically return erroneous results, that are only detected and corrected by a completely separate program.). See also, Haber, *supra* note 1175, at 17 (It has been observed that the checking of a crime lab technician's work by another technician or supervisor serves to ratify the formers conclusion(s), rather than to independently verify its soundness.)

[1251] "Scientific Methods," *supra* note 24, at 119-120. (Among the aspects of the scientific method are: "testable consequences;" "repetition of the test;" and "reliability and accuracy.").

[1252] More than being a mathematical expression, the simple expression that 2+2=4 is scientifically grounded as being verifiably true.

[1253] "Scientific Methods," *supra* note 24, at 121.

[1254] *Daubert*, 509 U.S. 579 (1993)

[1255] See e.g., Baden, *supra* note 46, at 234. ("In spite of *Daubert*, I see, hear and read about junk, bad and pathological science all the time.") *Id.*

[1256] This is the case whether or not the testimony is scientifically sound. In the realm of fingerprint analysis, it has been noted that "... jurors place great weight on the testimony of fingerprint experts, and rank fingerprint evidence as the most important scientific reason why they vote for conviction. Meagher (2002) could not remember an instance in which an FBI examiner made a positive identification and the jury set the identification aside and acquitted the defendant." Haber, *supra* note 60, at 6.

[1257] Quote from, Woodworth, *supra* note 6, at 48.

[1258] For a penetrating analysis of how a reliance of technology affects society as a whole and the attitudes of people in general, see, Jacques Ellul, *The Technological Society* (New York: Random House, 1964).

[1259] *Id.* at 4.

[1260] In the realm of fingerprint analysis, it has been noted that "... jurors place great weight on the testimony of fingerprint experts, and rank fingerprint evidence as the most important scientific reason why they vote for conviction. Meagher (2002) could not remember an instance in which an FBI examiner made a positive identification and the jury set the identification aside and acquitted the defendant." Haber, *supra* note 60, at 6.

[1261] See *supra* Chapter 3: Roll Call of Suspect Crime Labs and Expert Prosecution Witnesses, for numerous examples of a defendant exonerated after being convicted by jurors who relied on crime lab technician expert testimony.

[1262] *Supra* notes 50-89, and accompanying text related to crime lab proficiency test results.

[1263] See *supra* Chapter 3: Roll Call of Suspect Crime Labs and Expert Prosecution Witnesses, for numerous examples of a defendant exonerated after being convicted by jurors who relied on crime lab technician expert testimony.

[1264] *Supra* notes 992-1004 and the accompanying text.

[1265] To understand the corrupting effect of crime labs working hand in glove with police agencies and prosecutors, one need look no further than Lord Acton's oft cited observation about the corrupting nature of power. ("power tends to corrupt, and absolute power corrupts absolutely."). Lord Acton, *Letter from Lord Acton to Bishop Mandell Creighton* (Apr. 3, 1887), *in* 1 *The Life and Letters of Mandell Creighton* ch. 13 (Louise Creighton ed. 1904), available at http://www.bartleby.com/66/9/2709.html (last visited January 24, 2004). A crime lab literally has the power of convicting any person that it issues a negative report about or that a technician testifies is linked to a crime through the analysis of the prosecution's physical evidence.

[1266] One more observation to add to the many related throughout this document supporting this contention, is, "Incredibly, forensic scientists do not have to establish competence by obtaining a license or certification – even by their peers. There are no federal requirements and, to date, no state has demanded them." Kelly, *supra* note 25, at 22.

[1267] Sherrer, *supra* note 414, at 567.

[1268] For examples of this, see, *supra* Chapter 3. XIV: Louise Robbins - Forensic

Anthropologist; *supra* Chapter 3.XV: Michael West - Forensic Dentist; *supra* Chapter 3. XVI: Sandra Anderson – Cadaver Finding Dog Trainer; and *supra* Chapter 3.XVII: Anthony Pellicano – Audio Expert Extraordinaire.

[1269] An example of this is that in the wake of the Office of the Inspector General's report on the FBI crime lab issued in April 1997, nothing substantive has changed in its operation. See e.g., ("Far from being rectified, false testimony by FBI lab agents is still being presented in criminal trials around the country, influencing jurors and compromising trials to the point where it's difficult to determine the guilt or innocence of some defendants."). Fischer *supra* note 1009, at 113. This is due to a significant degree because of the absence of outside accountability in the bureaucratic operation of a crime lab. See e.g., Sherrer, *supra* note 114, at 258 ("Bureaucrats are protected by a nearly complete absence of outside accountability. They can do almost anything under the color of acting as a government employee without fear of legal consequences or personal financial accountability to anyone they harm.") *Id.*

[1270] Fischer *supra* note 1009, at 113. ("The FBI's leadership acknowledged the problems and assured Congress they had been fixed.") The OIG's investigation was in response to whistleblowing by Frederic Whitehurst. See e.g., Johnston, *supra* note 111.

[1271] In regards to the continuing pattern of conduct that inspired the OIG's investigation, see e.g., Fischer *supra* note 1009, at 113 ("Far from being rectified, false testimony by FBI lab agents is still being presented in criminal trials around the country, influencing jurors and compromising trials to the point where it's difficult to determine the guilt or innocence of some defendants.") *Id.*

[1272] *Id.* at 113.

[1273] Committee on Scientific Assessment, *supra* note 134.

[1274] Dan Eggen, "Study Faults FBI Bullet Tests," *The Washington Post*, Feb. 11, 2004, p. A12.

[1275] Committee on Scientific Assessment, *supra* note 134, at 93 (5-3).

[1276] Eggen, *supra* note 1274, at A12.

[1277] See e.g., Sherrer, *supra* note 114, at 261 (In summarizing what could be expected in regards to the influence of government bureaucracies – to which group crime labs belong – it is stated, "current developments give scant reason to expect a cessation of that growth anytime soon.") Also, *Id.* at 258 (There is no incentive for a bureaucracy to make any meaningful change, since, "Bureaucrats are protected by a nearly complete absence of outside accountability. They can do almost anything under the color of acting as a government employee without fear of legal consequences or personal financial accountability to anyone they harm.") *Id.*

[1278] See e.g., *supra* notes 50-89 and accompanying text. (Discusses the high error rates in the proficiency tests of crime lab technicians conducted during various years from 1974 to 2001). That high error rate could be reduced by double blind proficiency testing procedures. See, e.g., Haber, *supra* note 60, at 9. ("Research on double blind testing procedures … show that test score results are inflated when people know they are being tested, and when they and their supervisors know of the importance of the test results."), and, Haber, *supra* note 1175, at 3, 12-13, esp. 12 ("… double blind test procedures uncover significantly more errors than single blind…").

[1279] The FBI's crime lab has been resistant to the adoption of any type of quality controls over the results it produces. The one concession it has made, being accredited by the ASCLD in 1999, has been described as a meaningless "perfunctory exercise." Fischer, *supra* note 1009, at 149. Given their intimate association with police agencies and prosecutors, it would be incongruent for a crime lab to embrace the use of scientific double-blind testing techniques that could complicate their delivery of results consistent with the prosecution's theory of the case. This situation is summed up in the a former forensic lab technician's frank observation,

Menace To The Innocent

"People say we're tainted for the prosecution. Hell, that's what we do! We get our evidence and present it for the prosecution." Kelly, *supra* note 25, at 16. (Quoting an interview of the former technician with author Phillip Wearne in March 1997.)
[1280] One example that this is widely known within the law enforcement fraternity was the frank observation by a former forensic lab technician, "People say we're tainted for the prosecution. Hell, that's what we do! We get our evidence and present it for the prosecution." *Id.* at 16 (Quoting an interview with author Wearne in March 1997).
[1281] See, *supra* notes 1242-1247 and accompanying text.
[1282] In a federal criminal case the defendant currently has a right to discovery of scientific "results and reports" under *Federal Rules of Criminal Procedures* Rule 16. However there is no requirement that a report be written documenting any phase of the testing process, and as the authors of *Tainting Evidence* observed, "That is a loophole the FBI and other crime labs have proven adept at exploiting." Kelly, *supra* note 25, at 27.
[1283] For a discussion of the ease of creating fake fingerprints and photographs, which far exceeds the requirements of creating fake documents, see, *supra* notes 566-569 and accompanying text.
[1284] Such subterfugal tactics would be particularly practical if the defense did not have, or was not provided with the financial resources to have the source evidence independently evaluated. See e.g., Kelly, *supra* note 25, at 27 ("The vast majority of defendants in criminal courts in the United States do not have access to forensic expertise, even though they will almost certainly face forensic evidence from the prosecution...") Furthermore, simply reviewing the crime lab's documentation of a test(s), or retesting the same sample, would be much more likely to confirm its conclusion, than would an independent evaluation of the item(s) for its evidentiary value. See e.g., Thompson, *supra* note 10, at 48. (The more conscientious a retest is performed, the more likely it is to confirm the initial false positive attributable to contamination.)
[1285] The FBI and other crime labs have strenuously resisted substantive quality controls or outside oversight that could interfere with their pro-prosecution culture. *Supra* notes 1279-1281 and accompanying text. (It is not in the self-interest of crime labs to subject themselves to any limitation in the range of options they have available in how the examination of prosecution evidence is handled, and its probative value testified to in court.)
[1286] *In Re Winship*, 397 U.S. at 363-364 (1970).
[1287] That principle first enunced by René Descartes is known as methodic doubt. See, "Methodic Doubt," *The New Encyclopedia Britannica*, Vol. 8, 15th Ed., 2002, at 70.
[1288] *Id.* at 70.
[1289] *Id.* at 70.
[1290] This is either self-evidently true, or the purpose of a trial is not to ascertain a defendant's actual level of culpability or its absence, but rather, to use the officious atmosphere of the courtroom to effectively rubber-stamp the prosecution's theory of the alleged crime and inflict punishment on the defendant, whether or not it is "legally" warranted.
[1291] Quoted in "Methodic Doubt," *supra* note 1287.
[1292] See e.g., John McManus, "Cot Deaths and the Adversarial Justice System," January 22, 2004, article sent to author from MOJUK.org.uk email service, http://www.mojuk.org.uk/. John McManus is Project Co-ordinator of MOJO Scotland. ("Michael Mansfield stated in the Angela Canning appeal that "the jury had been forced to rely on expert witnesses who had given contradictory evidence." This happens all too often in cases of misjustice and leaves the jury in an unenviable position.")
[1293] See e.g., *Id.* ("[I]n our adversarial system, we are to quick to believe the views of prosecution experts based on expediency, tunnel vision or malfeasance, and the bottom line is that no-one is ever held responsible. Tragedy is followed by cover-ups, the logic of the ostrich,

and still it continues.") *Id.* Although this is specifically in reference to the situation in the United Kingdom, there are significant similarities between their evidentiary procedures and those in the U.S., and in fact, in June 2001 the U.K. adopted the FBI's no numerical standard for fingerprint examinations. Cole, *supra* note 1, at 286. (Britain abandoned the sixteen point standard in 2001.). In the realm of fingerprint analysis, it has been noted that "... jurors place great weight on the testimony of fingerprint experts, and rank fingerprint evidence as the most important scientific reason why they vote for conviction." Haber, *supra* note 60, at 6.

[1294] McManus, *supra* note 1292. McManus expresses the opinion in the article that the inquisitorial system predominating on the European Continent is more suited to arriving at the truthful scientific value of prosecution evidence.

[1295] The lack of a scientific basis for the process is demonstrated by considering a basic example. If the prosecution witness testified that $2,345 \times 3,456 = 8,104,230$, and a defense witness testified that no, it equals $8,104,320$, there is a fundamental disagreement that can conclusively be resolved to a scientific certainty by the mathematical fact that the prosecution witness is wrong, and the defense expert is right. Yet today, even under *Daubert*, a federal court judge or jury has the Alice in Wonderland option of rejecting the scientific basis of the defense witness' correct conclusion, while accepting the prosecution witness' mathematically false solution.

[1296] *Daubert*, 509 U.S. at 593 ("'Scientific methodology today is based on generating hypotheses and testing them to see if they can be falsified; indeed, this methodology is what distinguishes science from other fields of human inquiry.").

[1297] *Daubert*, 509 U.S. at 593

[1298] *Daubert*, 509 U.S. at 593 (Although the Court did not specifically state that in its ruling, it is implicit in the testing process it outlined as required to determine if testimony purporting to be scientific is in fact scientific.)

[1299] *Supra* notes 1203-1211 and accompanying text.

[1300] In McManus, *supra* note 1292, it is proposed that pathological forensic testimony can be reduced by adoption of the European Continents inquisitorial prosecution method.

[1301] The attitude that the function of crime labs is to present pro-prosecution evidence was perhaps most graphically expressed by a former lab technician's observation, "People say we're tainted for the prosecution. Hell, that's what we do! We get our evidence and present it for the prosecution." Kelly, *supra* note 25, at 16.

[1302] The very purpose of *methodic doubt* is to contribute to certainty of a statement's truthfulness, thus in its absence, doubt is an inevitable byproduct.

[1303] The court has a separate interest representing the principle of seeking to ensure justice, whereas the prosecution ostensibly represents the interests of the government, and the defense of the accused. Justice Breyer noted in *General Electric Co. v. Joiner*, 522 U.S. 136 (1997), that judges have the inherent authority to appoint experts apart from those retained by the prosecution and defense, and that since "Judges are not scientists," *Id.* at 148 they should make use of that power whenever it will contribute to a more reliable ascertainment of the truth. *Id.* at 149-150 (Justice Breyer concurring).

[1304] This cross-analysis process would also apply when an expert testified about evidence not adduced from a forensic laboratory, such as Sandra Andersen: She testified about cadaver parts found with the aid of her dog Eagle. See, *supra* Chapter 3.XVI: Sandra Anderson – Cadaver Finding Dog Trainer.

The application of *methodic doubt* is consistent with the the need for independent laboratory oversight was explained by FBI whistleblower Frederic Whitehurst. See, Martin, *supra* note 131. It was his revelations that led to an inspection of the FBI crime lab by the Office of the Inspector General, and the issuance of its report and recommendations in April 1997. In April 2003, after the latest round of FBI lab scandals broke in the press, Whitehurst

Menace To The Innocent

said, "the lab should be subject to independent regulation and inspection. ... It's a horribly huge deal. ... The problem is, they're telling you – the people that are paying for the upkeep of the lab – that you're not allowed to know what the problems are." *Id.*

[1305] *In re Winship*, 397 U.S. at 363-364. ("The reasonable-doubt standard plays a vital role in the American scheme of criminal procedure. It is a prime instrument for reducing the risk of convictions resting on factual error. The standard provides concrete substance for the presumption of innocence — that bedrock "axiomatic and elementary" principle whose "enforcement lies at the foundation of the administration of our criminal law."") *Id.*

[1306] Although this would admittedly be a radical innovation from the current situation that doesn't require any form of independent cross-ascertainment of the true probative value of the prosecution's evidence, it would involve minimal, and possibly no disruption of existing processes.

[1307] Cole, *supra* note 1, at 286. (Britain abandoned the sixteen-point standard in 2001.); and at 202 ("... a latent print comparison resulting in fewer than sixteen points of similarity would be declared inconclusive automatically.").

[1308] *Methodic Doubt, supra* note 1287, at 70.

[1309] That is why, at least until June 2001, in Britain "... a latent print comparison resulting in fewer than sixteen points of similarity would be declared inconclusive automatically." Quote in Cole, *supra* note 1, at 202.

[1310] See e.g., Haber, *supra* note 1175, at 19.

[1311] See e.g., *Id.* at 19, unpublished article available at, http://humanfactorsconsultants.com/research.html (last visited January 12, 2004). ("If errors are truly independent (as double blind assures), then the chances of that both examiners will make the same error is 0.20 x 0.20 = 0.04, or 4%--a far cry from 20%. A third independent verifier drops the chances that an error will be agreed upon by all three is 0.20 x 0.20 x 0.20 = 0.008, or less than one in a hundred.").

[1312] For example, in September 2004, 38 cases in Britain involving women convicted of killing one of her children were identified after a special review by Attorney General Lord Goldsmith, as possibly bieng unsafe due to "serious disagreement between distinguished and reputable experts," regarding the medical evidence underlying the conviction. Verkaik, *supra* note 969. That was about 14%, one out of seven cases reviewed. *Id.* Twenty-four of the cases were recommended for referal to the Criminal Cases Review Commission (CCRC) and 14 were recommended for appeal. *Id.* Critics suggested the review should be expanded to refer all cases to the CCRC that were based on the testimony of a single expert. *Id.* This mass review was prompted by the exoneration of Sally Clark in January 2003, after she was convicted in 1999 of murdering two of her infant children, and of Angela Canning in December 2003 after her 2002 conviction of murdering two of her children. In both cases, the children actually died of natural causes. McManus, *supra* note 1292.

[1313] *Id.*

[1314] See e.g., *Methodic Doubt, supra* note 1287, at 70. ("The hope is that, by eliminating all statements and types of knowledge the truth of which can be doubted in any way, one will find some indubitable certainties.").

[1315] With no apparent harm to the ability of law enforcement to function effectively, that was a consequence in this country for a panoply of Constitutional violations by police and prosecutors for about 200 years until the harmless error doctrine was expanded beginning in the early 1990s to make such violations subject to harmless error analysis. See e.g., Sherrer, *supra* note 414, at 565-570. Insofar as the analysis of of crime lab evidence is concerned, in Britain until June 2001, if a suspect's prints couldn't be matched to a latent crime related print on at least 16 points, the print was automatically deemed to be inconclusive. See e.g., Cole, *supra* note 1, at 202 ("... a latent print comparison resulting in fewer than sixteen points of similarity would be declared inconclusive automatically.") There is no substantive evidence that holding the police and prosecutors to a standard of producting evidence against an accused that is both credible and constitutionally (lawfully in Britain) obtained endangers the public.

[1316] *In re Winship*, 397 U.S. at 363-364. ("The reasonable-doubt standard plays a vital role in the American scheme of criminal procedure. It is a prime instrument for reducing the risk of convictions resting on factual error. The standard provides concrete substance for the presumption of innocence — that bedrock "axiomatic and elementary" principle whose "enforcement lies at the foundation of the administration of our criminal law."") *Id.*

[1317] *Fed. R. Evi.* 102.

[1318] *General Electric Co.*, 522 U.S. at 148. (Justice Breyer concurring).

[1319] *Id.* at 149-150 (Justice Breyer concurring).

[1320] In the realm of fingerprint analysis, it has been noted that "... jurors place great weight on the testimony of fingerprint experts, and rank fingerprint evidence as the most important scientific reason why they vote for conviction. Meagher (2002) could not remember an instance in which an FBI examiner made a positive identification and the jury set the identification aside and acquitted the defendant." Haber, *supra* note 60, at 6.

[1321] *Id.* A judge's bias to rely on prosecution expert testimony is highlighted by what happened in the Brandon Mayfield case in May 2004, when three FBI fingerprint experts, and one outside court expert determined that one of his fingerprints was on a plastic bag with detonators in it that was linked to the March 2004 train bombings in Madrid, Spain that killed 191 people. See Hans Sherrer, "That's Not My Fingerprint, Your Honor," *Justice Denied*, Issue 25, Summer 2004, 11-14, 19. It was later conclusively proven, and the FBI admitted, that Mayfield's fingerprint was not on the plastic bag. *Id.*

[1322] The use of a triad of independent experts representing the three interests directly involved in a criminal case who would need to unanimously agree to the probative value of the prosecution's evidence before it could be presented as "true," is an expression of Rationalist René Descartes understanding of the difficulties in relying on normal channels of truth seeking: "He found knowledge from tradition to be dubitable because authorities disagree; empirical knowledge dubitable because of illusions, hallucinations, and dreams; and mathematical knowledge dubitable because people make errors in calculating." Quote from, "Methodic Doubt," *supra* note 1287, at 70.

[1323] See e.g., Herma Silverstein, *Threads of Evidence* (Connecticut: Twenty-First Century Books, 1996), at 8.

[1324] *Id.* at 8.

[1325] *Id.* at 8.

[1326] *Id.* at 8.

[1327] *Id.* at 10.

[1328] *Id.* at 5. ("Until the 1960s, the police did not fully utilize forensics, and preferred "good old detecting.")

[1329] *Id.* at 5.

[1330] *Id.* at 6. (citation omitted).

[1331] The importance of experts to the process is precisely why it is important to institute an iron-clad safeguard, such as methodic doubt, against the use of "evidence we can't see," when its evidentiary value is not unanimously ascribed to by independent experts. See *supra* Chapter 18 – Methodic Doubt Can Overcome Pathological Science In The Courtroom.

[1332] Richard Harris, *The Fear of Crime* (New York: Frederick A. Praeger 1969), at 23-24.

[1333] *Id.*

[1334] See e.g., Silverstein, *supra* note 1323, at 5. ("Until the 1960s, the police did not fully utilize forensics, and preferred "good old detecting."", and "Gone are the days when we'd go to a crime scene and pick up whatever we could see. Nowadays we're more interested in evidence we can't see." *Id.* at 6.

[1335] Harris, *supra* note 1332, at 23-24.

[1336] Silverstein, *supra* note 1323, at 8 ("The first American crime lab opened in Los Angeles, California, in 1923.").

[1337] *In re Winship*, 397 U.S. at 363-364 (Proof beyond a reasonable doubt requires the prosecution to meet its burden of proof to sustain a person's conviction of a crime.).

[1338] See e.g., Underwood, *supra* note 4, at 167 ("It is no secret that expert witnesses can be "co-opted" by the prosecution-they may be little more than hired guns of the state."), referencing note 87 that cites, Thompson, *supra* note 809, at 1115.

[1339] See e.g., Kelly, *supra* note 25, at 232-233. (The LAPD and its crime lab's lack of cleanliness standards and practices that create the possibility of contamination during transportation, handling, storage and testing of evidence was a central issue in the O.J. Simpson case. The unaddressed issue remains important in state crime labs as well as the FBI's. The deficient procedures used to collect, handle, transport, test and store crime scene evidence examined by a forensic lab is best illustrated by contrasting it with the clean room procedures implemented by computer product manufacturers to avoid contamination of chips, motherboards, etc. At a minimum technicians in a computer chip manufacturing facility wear masks, gloves and smocks (similar to the garb medical personnel wear during an operation) to avoid contaminating what they are working on – which could affect its operation. Yet crime crime related evidence is typically handled, worked on and stored in markedly unsterile conditions that heighten the likelihood it could be affected by contaminants of varying sorts. See e.g., Cole, *supra* note 1, at 300. ("They attacked the weakest link in the processing of the DNA evidence, the work of the LAPD's forensic technicians Dennis Fung and Andrea Mazzola, who had committed numerous procedural errors in recovering, storing, and transporting the evidence to the LAPD crime laboratory. ... In his closing argument Scheck compared the hygiene of the LAPD's forensic evidence truck to a New York City restaurant in which cockroaches are visible, and he suggested that the evidence implicating Simpson had been contaminated or, worse, planted by the police themselves.") *Id.*

[1340] Kelly, *supra* note 25, at 21. ("Traditionally, many FBI forensic scientists have not used protocols – the recipes for analyses and the touchstones of scientific procedure – despite the fact that all scientists accept that not using them produces only experimental, not proven, outcomes. Indeed, in some crime labs, established protocols do not even exist.") *Id.*

[1341] See e.g., *Id.* 20-21. ("Documentation is a case in point. Examiners have proven remarkably loath to write up their bench notes in any adequate scientific manner. No names, no chain of custody history, no testing chronology, no details of supervisory oversight, no confirmatory tests, no signatures ... Since lab reports are "discoverable" and have to be handed to the defense, the FBI lab believes that as little as possible should be given away.") *Id.*

[1342] See e.g., Baden, *supra* note 46, at 233 ("I would also include dressing up speculation in scientific terms to make things that are not proven sound as though they are. When people do that they are giving the appearance of science when the work isn't scientific. To me junk science is wishful thinking wrapped in a pseudoscientific cloak.") *Id.*

[1343] See e.g., *Id.* at 233 ("The Union of Concerned Scientists in Cambridge, Massachusetts which may have coined the phrase, defines junk science as, "work presented as valid science that falls outside the rigors of the scientific method and the peer review process."") *Id.*

[1344] The adoption of procedures consistent with methodic doubt would resolve the problem of a lack of corroboration. See *supra* Chapter 18: Methodic Doubt Can Overcome Pathological Science In The Courtroom.

[1345] See e.g., Kelly, *supra* note 25, at 22. ("Incredibly, forensic scientists do not have to

establish competence by obtaining a license or certification – even from their peers.") *Id.*
[1346] See e.g., *Id.* at 22. ("In the cauldron of the courtroom, testifying beyond one's expertise becomes common... When only one expert is appearing in a multidiscipline case, ... it's also tempting for examiners to embellish, exaggerate, or even lie about their credentials.") *Id.* See also, "It's an old boys' network, said William C. Thompson, criminology and law professor at the University of California-Irvine. "It's the absolute bare bones that's needed to run a lab. It isn't the best scientific work that can be done. The labs have manufactured credentials for themselves. If you have people who are willing to manufacture credentials, what else are they making up?" Teichroeb, *supra* note 281.
[1347] See e.g., Thompson, *supra* note 10, at 47, 53. The testing of forensic technicians is neither external (i.e., administered and evaluated by an organization unassociated with the laboratory whose technicians are tested), nor is it blind, nor is it representative of the real-life situations requiring analysis in a typical criminal case. For a contrast to clinical labs, see, Kelly, *supra* note 25, at esp 29-31. (The performance of technicians in clinical laboratories are evaluated by blind proficiency tests.)
[1348] The attitude that the function of crime labs is to present pro-prosecution evidence was perhaps most graphically expressed by former lab technician, "People say we're tainted for the prosecution. Hell, that's what we do! We get our evidence and present it for the prosecution." *Id.* at 16. See also, *supra* notes 969-981 and accompanying text.
[1349] In reference to Washington State Police Crime Lab technician Arnold Melnikoff's skill level, Seattle University Law Professor John Strait observed, his 30% error rate in analyzing prosecution evidence, "wouldn't pass a first-year college chemistry class." Teichroeb, *supra* note 261, at B5. A triad of experts based on the principles of methodic doubt can be expected to have detected, and barred the prosecution's use of Melnikoff's insubstantial test results agaisnt a defendant. *Supra* notes 1282-1306 and accompanying text.
[1350] As closely allied in the law enforcement system, police agencies and prosecutors are members of what has been referred to as the bureaucratic brotherhood. See, Sherrer, *supra* note 114, at esp. 256-260. One feature of bureaucratic alliances is the people involved tend to see their personal self-interest as being indistinguishable from that of the bureaucracy they are a part of. *Id.*
This also applies to any government funded laboratory that tests crime related physical evidence in a case.
[1351] See e.g., Underwood, *supra* note 4, at 167 ("It is no secret that expert witnesses can be "co-opted" by the prosecution-they may be little more than hired guns of the state."), referencing note 87 that cites, Thompson, *supra* note 809, at 1115.
[1352] The attitude that the function of crime labs is to present pro-prosecution evidence was perhaps most graphically expressed by a former lab technician, "People say we're tainted for the prosecution. Hell, that's what we do! We get our evidence and present it for the prosecution." Kelly, *supra* note 25, at 16. See also, *supra* notes 963-976 and accompanying text.
[1353] In the realm of fingerprint analysis, it has been noted that "... jurors place great weight on the testimony of fingerprint experts, and rank fingerprint evidence as the most important scientific reason why they vote for conviction. Meagher (2002) could not remember an instance in which an FBI examiner made a positive identification and the jury set the identification aside and acquitted the defendant." Haber, *supra* note 60, at 6.
[1354] In regards to various aspects of fingerprint analysis that are suspect or known to be insubstantial see, *supra* Chapter 6: Fingerprint Analysis: Voodoo Palmed Off As Science.
[1355] The case of Lloyd Miler is illustrative of this. He was wrongly convicted of murdering a young girl and sentenced to death based on false expert prosecution testimony that a piece of clothing found near the crime scene had his blood on it. In fact it was neither his clothing nor

did it have his blood on it. See e.g., Bedau, *supra* note 25, at 141-152.

[1356] In theory, a central function of the judicial process is to ensure a person is only convicted of a crime after being proven guilty beyond a reasonable doubt. See, *In re Winship*, 397 U.S. at 363-364.

[1357] Silverstein, *supra* note 1323, at 5. ("Until the 1960s, the police did not fully utilize forensics, and preferred "good old detecting.") *Id.*

[1358] See e.g., Baden, *supra* note 46, at 233 ("Junk science is not new."). See also, Silverstein, *supra* note 1323, at 5. "Until the 1960s, the police did not fully utilize forensics, and preferred "good old detecting."", and "Gone are the days when we'd go to a crime scene and pick up whatever we could see. Nowadays we're more interested in evidence we can't see." *Id.* at 6.

[1359] Viewed in the most favorable light, *Daubert* and its progeny may have been well meaning attempts to correct this situation. However they have been ineffective at doing so, since there has not been an increased level of certainty that expert testimony in a particular case is reliable. What those cases have done however, is contribute to the aura of scientificness surrounding expert prosecution testimony, which has the practical effect of further eroding the presumption of innocence standard. See e.g., *supra* notes 1246-1250 and accompanying text.

[1360] Due to their association with the law enforcement system's bureaucratic structure, prosecution expert witnesses are inexorably tied first and foremost to serve the interests of that bureaucracy. See e.g., Sherrer, *supra* note 114, at 258-260. Whistleblowers such as former FBI lab technician Fredrick Whitehurst are a rare exception to that rule, however they pay the price of being fired or otherwise ostracized. See e.g., Johnston, *supra* note 111. See also, the FBI has continued J. Edgar Hoover's policy of not tolerating "… any whistleblowers in the FBI." Swearington, *supra* note 114, at 55.

[1361] See e.g., Michael Oliver Foley, "Police Perjury: A Factorial Survey," *U.S. Dept of Justice, National Institute of Justice* (2000). (The article abstract summarizes the findings of the survey: "A literature review revealed that lying is as common or more common than honesty in modern life. The courts, police agencies, and society have acknowledged, justified, and approved the use of lying and deception by police."). See also, several articles written by Margaret L. Paris, a University of Oregon Law School professor concerning issues related to law enforcement lying: Margaret L. Paris, *Lying to Ourselves*, 76 Or. L. Rev. 817 (1997); Margaret L. Paris, "Trust, Lies, and Interrogations," 3 *Va. J. Soc. Policy & Law* 3 (1995); and, Margaret L. Paris and Julie Armstrong, "Who Can We Trust If Not the Police?," *Seattle Post-Intelligencer*, Oct. 24, 1995, at A9.

[1362] *Id.*

[1363] *Id.* The general dishonesty of law enforcement personnel is indicative of the pervasiveness of that situation throughout U.S. society. See e.g., David Callahan, *The Cheating Culture: Why More Americans Are Doing Wrong to Get Ahead* (San Diego: Harcourt 2004).

[1364] *Daubert*, and its progeny, however well meaning of an attempt they may have been to do so, have not increased the level of certainty that "expert testimony" in a particular case is reliable. See e.g., *supra* notes 1246-1250 and accompanying text. This includes testimony from freelance prosecution experts such as Sandra Anderson and Anthony Pellicano. See *supra* Chapter 3.XVI and Part3.XVII respectively.

[1365] See e.g., Baden, *supra* note 46, at 234. ("In spite of *Daubert*, I see, hear and read about junk, bad and pathological science all the time.")

[1366] Of the many such examples illustrating this that could be cited, see e.g., Editorial ("The FBI's flawed lab"), *supra* note 2. (The FBI crime lab engages in "shoddy work, withholding of relevant evidence from defense attorneys and outright bias in favor of prosecutions."); and, Kelly, *supra* note 25, at 16. (During an interview with author Phillip Wearne in March 1997, a former forensic lab technician was brutally frank in observing, "People say we're tainted for the prosecution. Hell, that's what we do! We get our evidence and present it for the prosecution.") *Id.* See also, *supra* notes 963-976 and accompanying text.

[1367] Also see *supra* notes 1287-1311 and accompanying text. It would be consistent for the same principle of indubitable certainty to be applied to expert testimony in civil cases. See, "Methodic Doubt," *supra* note 1287, at 70.

[1368] For a full explanation of this, see, *supra* Chapter 18: Methodic Doubt Can Overcome Pathological Science In The Courtroom.

[1369] Indubitable certainty of the evidence's probative value is necessary to overcome methodic doubt. See e.g., *supra* notes 1287-1321 and accompanying text.

[1370] The proposed solution is unobtrusive in the sense that there is nothing inherent in it that would be an impediment to its implementation. Any difficulties would arise form the opposition of police, prosecutors and judges to an impairment of their respective abilities to support, present and/or or authorize the courtroom use of pseudo-scientific or junk science evidence masquerading as scientific. For a full explanation of this, see, *supra* Chapter 18: Methodic Doubt Can Overcome Pathological Science In The Courtroom.

[1371] *The Innocents Database, supra* note 3, includes many accounts of innocent people victimized by insubstantial expert testimony.

[1372] Evidence that can't be seen and/or that is considered to require special training to understand is what is analyzed and testified to by an expert. See e.g., Silverstein, *supra* note 1323, at 6. (citation omitted). ("Gone are the days when we'd go to a crime scene and pick up whatever we could see. Nowadays we're more interested in evidence we can't see."). *Id.*

[1373] Lefcourt, *supra* note 2. ("It is the job of the jury to decide guilt or innocence, not a "13th juror" demonstrating bias in favor of the prosecution.").

[1374] For examples of the many suspect tactics prosecutors use to secure a conviction "at all costs," see, Sherrer, *supra* note 34; see also, "Prosecutorial Lawlessness," *supra* note 34.

[1375] See e.g., Underwood, *supra* note 4, at 167 ("It is no secret that expert witnesses can be "co-opted" by the prosecution-they may be little more than hired guns of the state."). See also, *supra* notes 991-1004 and accompanying text.

[1376] Justice Breyer observed "Judges are not scientists." *General Electric Co.,* 522 U.S. at 148. Neither judges nor prosecutors are required to have any scientific training or expertise, so it is somewhat incongruent for them to make evaluations and decisions related to determining the scientific validity of scientific tests and/or expert testimony concerning its possible evidentiary value. As explained in, *supra* Chapter 19: Crime Labs Are a 20th Century Invention That Contribute To Shortshrifting Reasonable Doubt, the veracity of a crime lab's test of evidence, and a technician's testimony related to it, is automatically suspect.

[1377] *The Innocents Database, supra* note 3, includes many accounts of innocent people victimized by insubstantial expert testimony.

[1378] Kelly, *supra* note 25, at 316.

[1379] This would not empower scientists with any revolutionary sway over the judicial process. In the present scheme, once a judge rules allegedly scientific evidence is admissible – it is assumed to have a scientific basis. Applying principles of indubitable certainty will simply help to ensure that such evidence in fact has a scientific basis – which would likely result in the exclusion of insubstantial evidence that would otherwise be presented to jurors.

The short answer to naysayers to the suggesting that the fatal defects in the crime lab model means they must be dissolved, and that the concerns of a triad of independent scientists must be respected for there to be an indubitable certainty of the reliability of expert testimony, is that they are mired in the idea that unreliable and insubstantial testimony pawned off as expert is a permanent and uncorrectable feature of this country's legal process. Those naysayers are only correct to the degree that idea is clung to – and that it will collapse like a house of cards when a critical mass of influential people understand it has no foundation except that belief.

www.ingramcontent.com/pod-product-compliance
Lightning Source LLC
Chambersburg PA
CBHW070459200326
41519CB00013B/2644